孟繁增　主审

Industrial Robot Technology Foundation and Application

工业机器人技术基础及应用

（英汉双语）

主　编　吕栋腾
副主编　胡春龙　王　刚　胥航军
参　编　黄　旭　常丽园　韩克华

扫码领取
- 本书配套视频
- 机器人编程课
- 机器人资讯台
- 机器人交流圈

西北大学出版社
·西安·

图书在版编目(CIP)数据

工业机器人技术基础及应用：英、汉/吕栋腾主编.
—西安：西北大学出版社，2022.4
ISBN 978-7-5604-4927-2

Ⅰ.①工… Ⅱ.①吕… Ⅲ.①工业机器人—教材—汉、英 Ⅳ.①TP242.2

中国版本图书馆CIP数据核字(2022)第067425号

工业机器人技术基础及应用(英汉双语)

GONGYE JIQIREN JISHU JICHU JI YINGYONG(YINGHAN SHUANGYU)

主编 吕栋腾

出版发行 西北大学出版社
(西北大学校内 邮编:710069 电话:029-88302621 88303593)
http://nwupress.nwu.edu.cn E-mail:xdpress@nwu.edu.cn

经　　销 全国新华书店
印　　装 西安市久盛印务有限责任公司
开　　本 787毫米×1092毫米 1/16
印　　张 18.75

版　　次 2022年4月第1版
印　　次 2022年4月第1次印刷
字　　数 553千字

书　　号 ISBN 978-7-5604-4927-2
定　　价 79.00元

PREFACE

前　言

机器人是“制造业皇冠顶端的明珠”，其研发、制造、应用是衡量一个国家科技创新和高端装备制造业水平的重要标志。随着科技水平的迅速提升，制造业智能化飞速发展，工业机器人作为一种柔性、高效、可靠、可直接执行动作的可编程自动化设备，是企业进行生产自动化改造、实现转型升级的最佳选择之一。大力发展工业机器人产业，对于打造制造业新优势，推动工业转型升级，加快制造强国建设具有重要意义。

本书根据高等职业教育能力培养目标的要求，“一带一路”职业教育国际交流与合作，按照国家“三教”改革的指导思想，考虑到技术发展和就业岗位，在兼顾学习内容深度和广度的同时，从知识应用和技能培养的实际出发，结合工业机器人技术最新教学实践，采用中英双语编写而成。

本书依托中国与巴基斯坦职业教育国际交流合作项目，采用中英双语和理实结合的教学方式。以 ABB 工业机器人为教学载体，内容主要包括：工业机器人的发展、应用及基础知识；工业机器人的数学模型建立、坐标系建立；工业机器人的机械、动力、感知系统；工业机器人的系统参数设置、手动操作和编程控制方法；工业机器人人才培养中心的建立；工业机器人系统建模与仿真等。介绍了工业机器人在复杂工艺流程的操作与编程方法，讲解了工业机器人系统设计的主要思路。“互联网+”思维有机融入，书中每一讲均可扫描二维码观看教学视频，配备了大量的立体化教学资源，学习者还可登录中文联盟平台进行在线自主学习。力图使学习者在修完全部内容后，能获得作为生产第一线的技术、管理、维护和运行技术人员所必须掌握的工业机器人操作、编程和系统维护等方面的基本知识和基本技能。

本书既可作为大中专院校工业机器人技术、机电一体化技术、电气自动化、智能控制技术、机械制造与自动化、数控技术等专业的教材，也可供相关工程技术人

PREFACE

员参考使用。

本书由陕西国防工业职业技术学院吕栋腾（11-2—15-6）担任主编，胡春龙（1-1—4-1）、王刚（4-2—6-2）、胥航军（7-1—7-2）担任副主编，陕西国防工业职业技术学院黄旭（8-1—8-2）、常丽园（9-1—10-2）、中国兵器工业213研究所韩克华（11-1）担任参编，孟繁增担任主审并对书中内容做最终的审读工作。

本书在编写过程中参考和引用了大量的资料和文献，在此谨向相关作者表示衷心感谢。由于编者水平有限，书中疏漏在所难免，恳请广大读者和专家批评指正。

编　者

CONTENTS

目　录

CONTENTS

CONTENTS

CONTENTS

CONTENTS

CONTENTS

Lecture 1-1

Introduction to Robot

微信扫一扫

教学视频

1-1 机器人简介

这是工业机器人工作站使用的一个典型例子。如图 1-1 所示。今天，机器人技术在很多不同的领域都有应用，有和外科医生一起为人类做手术的（图 1-2），帮助人们搬运重物的，娱乐领域的（图 1-3）等许多不同领域的机器人。

This is a typical example of the use of industrial robot workstation. As it is shown in Fig. 1-1. Today, robotics has been applied in many different areas, from robots working with a surgeon（Fig. 1-2）to operate on a human, to assisting workers, to carry a heavy load or robots in entertainment(Fig. 1-3), to many different fields.

Fig.1-1 Robot Workstation

图 1-1 机器人工作站

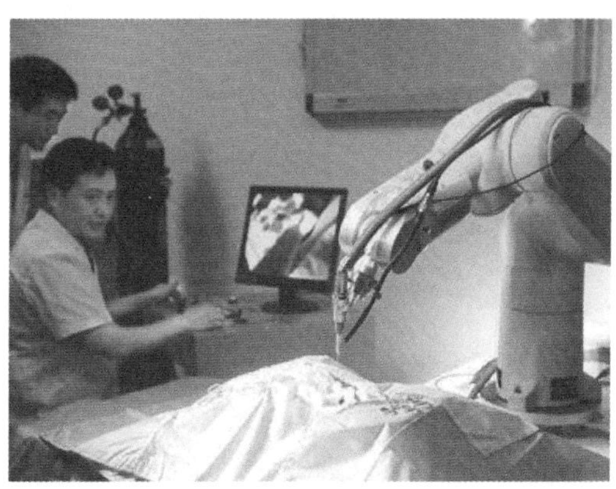

Fig.1-2 Robots Working With A Surgeon

图 1-2 医疗机器人

Fig.1-3 Interaction With The Computer
图 1-3 人机交互系统

We are able now to use all of these models and techniques to model human and to create a digital model of human that can be simulated and controlled to reproduce actual behavior captures from motion capture devices about human behavior.

The most exciting area in robotics is reproducing devices robots that look like human and behave like animals or humans. As it is shown in Fig.1-4. You could see Asimo here, which is the latest in a series of development. We also have different kinds of robots shown in Fig.1-5 to Fig.1-10.

我们现在能使用所有这些模型和技术模拟人类。可创建一个数字化的动作捕捉设备模拟人类的实际行为。

在机器人技术中，最令人兴奋的领域是制造出看起来像人类、行为像动物或人类的机器人。如图 1-4 所示。这是阿西木机器人，已有一系列产品。还有许多不同类型的机器人如图 1-5 到 1-10 所示。

Fig.1-4 Asimo Robot
图 1-4 阿西木仿生机器人

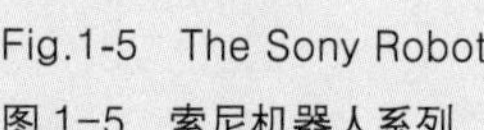
Fig.1-5 The Sony Robot
图 1-5 索尼机器人系列

Fig.1-6 HPR serial Robot in South Korea
图 1-6 韩国的 HPR 系列机器人

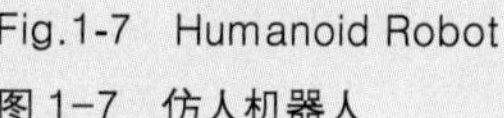
Fig.1-7 Humanoid Robot
图 1-7 仿人机器人

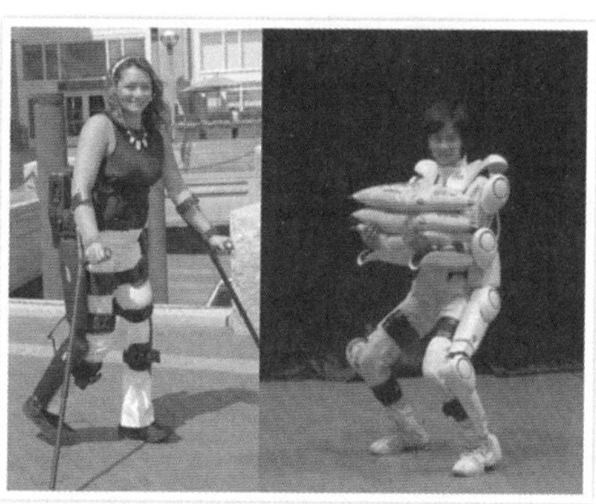

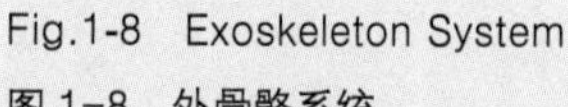
Fig.1-8 Exoskeleton System
图 1-8 外骨骼系统

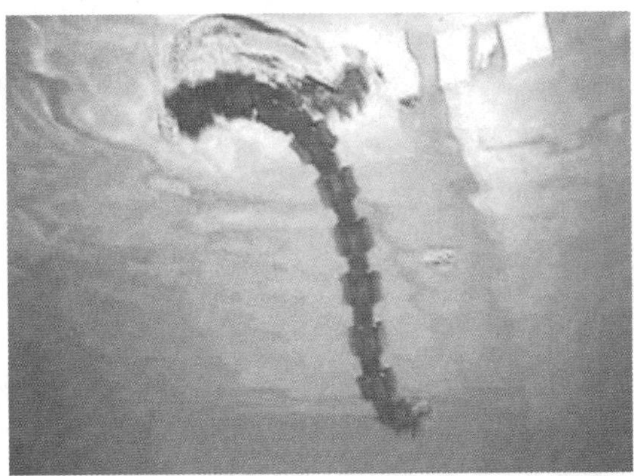

Fig.1-9 A Swimming Robot
图 1-9 游泳机器人

Fig.1-10 Robot Workstation
图 1-10 机器人工作站

The industrial robots in the enterprises often designs based on functional requirements so that we can acquire lower cost and more practicable.

企业中的工业机器人往往根据功能需求进行设计，以获得更低的成本和更大的实用性。

微信扫一扫

教学视频

Lecture 1-2
Overview of Industrial Robots

1-2 工业机器人概述

现在的工业机器人生产主要分布在日本、欧洲这两个地区。目前在工业领域中应用比较广泛的产品系列主要有ABB、FANUC、KUKA、Yaskawa等。如图1-11所示是不同厂家生产的工业机器人。

Most of today's industrial robots are built in Japan and Europe, major robotics companies include ABB (ABB), FANUC, KUKA Robotics, and Yaskawa Electric. Fig. 1-11 shows industrial robots produced by different manufacturers.

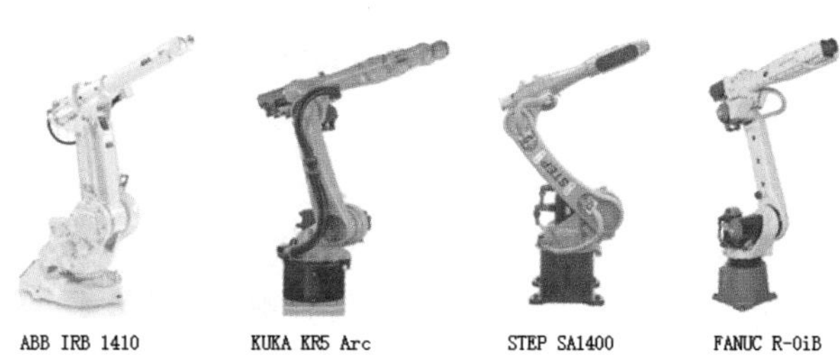

Fig.1-11 Industrial Robots From Different Manufacturers

图1-11 不同厂家生产的工业机器人

1962年，美国制造了第一个工业机器人，如图1-12所示。1969年，维克多·沙因曼在斯坦福大学发明了斯坦福手臂机器人，如图1-13所示。

The world's first industrial robot was brought to life in the United States in 1962 As it is shown in Fig. 1-12. In 1969, Victor Scheinman at Stanford University invented the Stanford arm. As it is shown in Fig. 1-13.

Fig.1-12 The First Industrial Robot

图1-12 第一台工业机器人

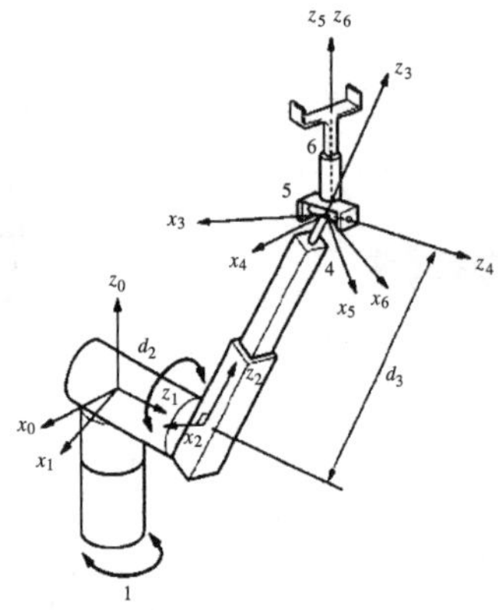

Fig.1-13 Stanford Arm

图 1-13 斯坦福手臂机器人

In 1978, American company Unimation introduces the industrial robot (Programmable Universal Machine for Assembly, PUMA), As it is shown in Fig.1-14. The use of universal automobile assembly line of PUMA robot marked the progress and the mature of the industrial robot. PUMA remains widely used it in the forefront of production in the enterprises today.

1978 年，美国 Unimation 公司推出通用工业机器人，如图 1-14 所示。应用于通用汽车装配线，这标志着工业机器人技术已经完全成熟。PUMA 至今仍然工作在工厂第一线。

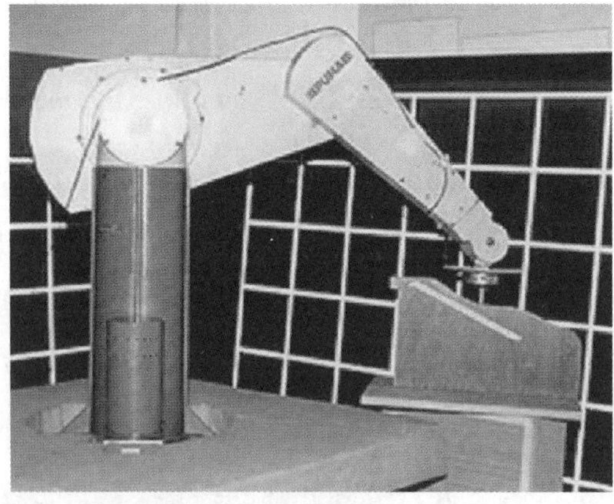

Fig.1-14 PUMA

图 1-14 可编程通用装配机

Industrial robotics took off quite quickly in Europe, with both ABB Robotics and KUKA Robotics bringing robots to the market in 1973.

随后工业机器人在欧洲大地快速发展，在 1973 年建立了 ABB、KUKA 两个工业机器人企业。

In 1985, The German company KUKA developed a new Z-shaped robotic arm that was designed to do away with the traditional parallelogram shape. The Z-

1985 年，德国库卡公司(KUKA)开发出一款新的 Z 形机器人手臂，它的设计摒弃了传统的平行四边形造型。该 Z 形机器人手臂可实现 3 个平移运动和 3 个旋转运动共 6 个自由度的运动维度，可大大节省制造工厂的场地空间。

2005 年，日本的 Motoman 推出了手臂大小的单臂（7 轴）和双臂（13 轴）机器人，所有的电源线都隐藏在机器人手臂中。如图 1-15 所示。

shaped robot arm can achieve three translational motions and three rotational motions with a total of six degrees of freedom, which can greatly save the site space of the manufacturing plant.

In 2005, Motoman, Japan, launched human sized single armed (7 axis) and dual armed robot (13 axis) with all of the supply cables hidden in the robot arm. As it is shown in Fig.1-15.

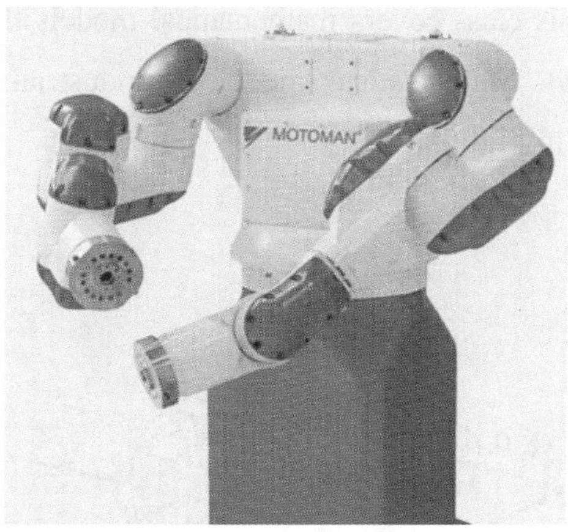

Fig.1-15 Dual Armed Robot
图 1-15 双臂机器人

1998 年，瑞典 ABB 公司开发出灵手（FlexPicke）机器人，它是当时世界上速度最快的采摘机器人。

In 1998, ABB, Sweden, launches the FlexPicker, the world's fastest picking robot.

微信扫一扫

教学视频

Lecture 2-1

Introduction to the Mathematical Model of Industrial Robot

2-1 工业机器人的数学模型简介

This class covers mathematical models that are essential. Mathematical model of industrial robot shown in Fig. 2-1.

本节课涵盖了最基本的数学模型。如图 2-1 所示为工业机器人数学模型。

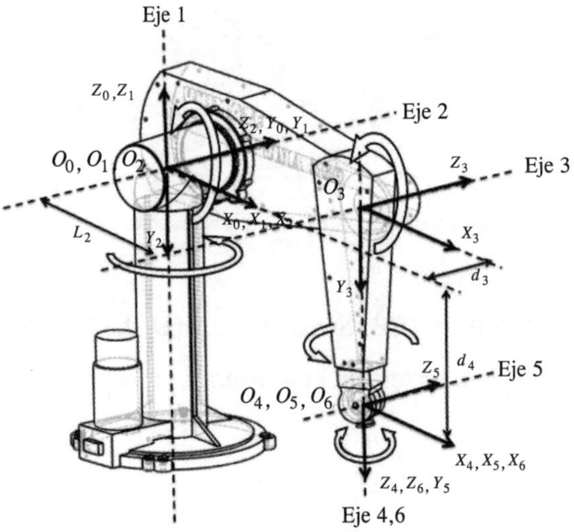

Fig.2-1 Mathematical Model of Industrial Robot

图 2-1 工业机器人数学模型

If we use a model to Simuiate the movement of the system. We can find many different ways to control the robot.First of all, you need to know where the robot is. And to know where the robot is. you need some sensors. We would need about 100 sensors if you have 100 degrees of freedom and that would be too expensive.

如果用一个模型来模拟系统的运动，可以发现有许多不同的方式来控制机器人。首先，你需要知道机器人在哪里。要知道机器人在哪里，你需要一些传感器。如果你有 100 个自由度，我们需要大约 100 个传感器，但这种做法成本过高。

编码器可以测量一个自由度，例如角度。这样你就会得到一个相对运动位置。你需要找到所有这些可见部件之间的关系，这样一旦机器人位置确定了，你就知道左臂的位置，右臂的位置，基座的位置。

其实我们可以通过一个数学模型来解决这个问题，运动学模型如图 2-2 所示。通过这个模型，可以对机器人进行定位，机器人各部件之间还可以形成一个联系。这就是数学模型教学的主要任务之一。

Encoders measure one degree of freedom, just like the angle. Now, that will give you the relative position. You need to find the relationship between all these vision bodies so that once the robot is standing, you know where the left arm is positioned and where the right arm is positioned, where the base is positioned.

We can solve problems by a mathematical model. The kinematics model is shown in Fig. 2-2. Using these models, you will locate the robot and have a coupling between these rigid bodies that are connected. This is one of the main tasks of the teaching of mathematical model.

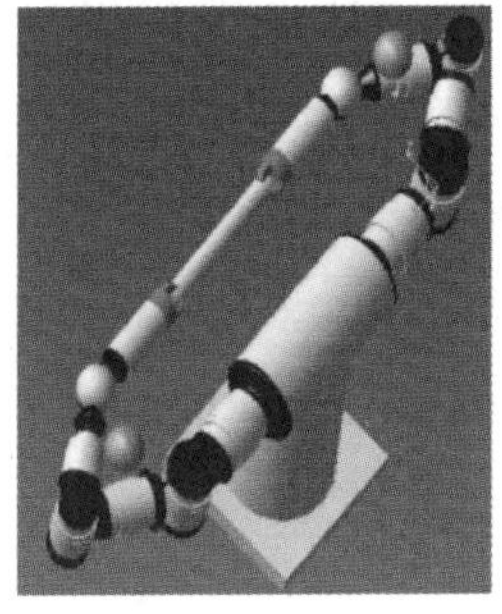

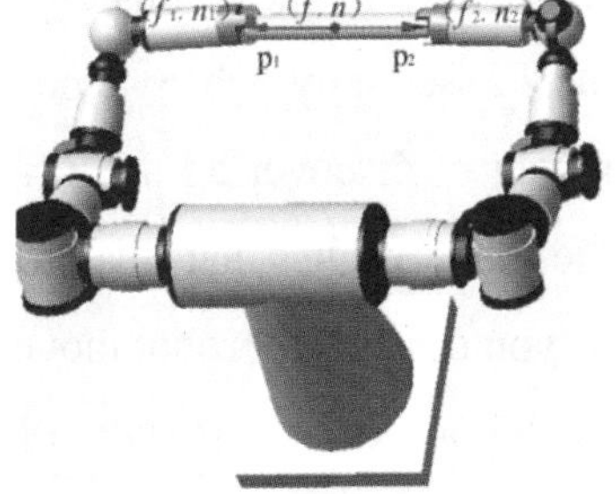

Fig.2-2 The Kinematics Model of Robot

图 2-2 机器人运动控制模型

微信扫一扫

教学视频

Lecture 2-2

Introduction to the Robotic Kinematics

2-2 机器人运动学介绍

The robotic kinematics include the forward kinematics and the inverse kinematics. The forward kinematics gives you the location of the hand. The inverse kinematics gives you given position for the hand that you desire.

Let's look at it in detail. The problem of moving this robot from one location to another. Suppose you would like to move this right arm. How do we do that? Essentially, what we need to do is somehow find a way of discovering a configuration through which the robot reaches that final goal position. If you solve it, you can control robot movement.The mathematical model of the robot is shown in Fig. 2-3.

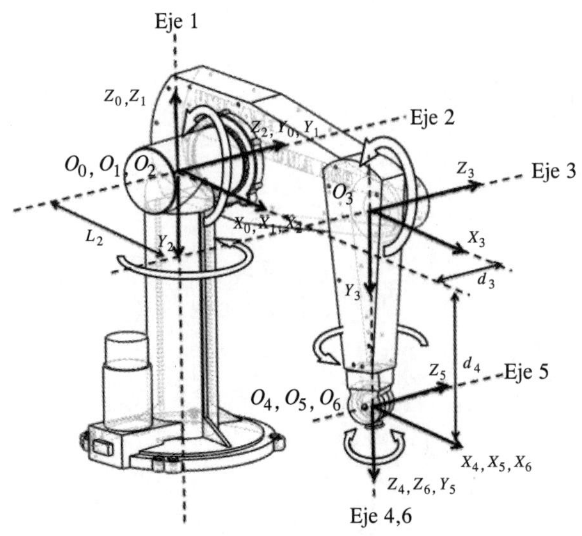

机器人运动学分为正向运动学和逆向运动学。正向运动学给出了手的位置。逆向运动学可以决定你希望手到达的位置。

让我们看个具体问题，把这个机器人从一个地方移动到另一个地方的问题。假设你想移动这个可移动的机械手平台，你如何实现呢？本质上，我们需要做的是找到一种方法来发现机器人到达最终目标位置的位型。如果你解决了这个问题，你就可以控制机器人运动了。机器人的数学模型如图 2-3 所示。

Fig.2-3 The Mathematic Model of Robotic

图 2-3 机器人数学模型

我们需要处理附加在这些不同物体上的坐标系之间的变换。这里，要知道这个末端执行器在哪里，最后你要知道要变换描述来确定末端执行器的位置。你需要在两个对象上的不同坐标系之间进行转换。

这里的机械臂是由一个固定的刚性物体，也就是基座，和另一个运动的刚性物体，我们称之为末端执行器组成。在这两个物体之间，所有的连杆将用来把末端执行器移动到某个位置。

我们将提出不同种类的关节和连杆，进而通过一组参数来描述连杆链条。在 20 世纪 70 年代早期，Denavit 和 Hartenberg 思考了这个问题，他们提出了一组参数来表示链上两个连续连杆之间的关系。如图 2-4 所示。

We need to deal with the transformation between frames attached to these different objects. Here, to know where this end factor is, you need to know you transform the descriptions to find, finally, the position of your end factor. You need transformations between different frames attached to both objects.

The mechanism that is the arm in this case is defined by a rigid object which is fixed, which is the base, and another rigid object that is moving, which we call the end factor. Between these two objects, you have all the links that are going to carry the end factor to move it to some location.

We will see that we are raising different kinds of joints, through those descriptions, we will describe the link and then we can describe the chain of links connected through a set of parameters. Denavit and Hartenberg in the early 70s thought about this problem, and they came up with a set of parameters to represent the relationship between two successive links on a chain. As it is shown in Fig. 2-4.

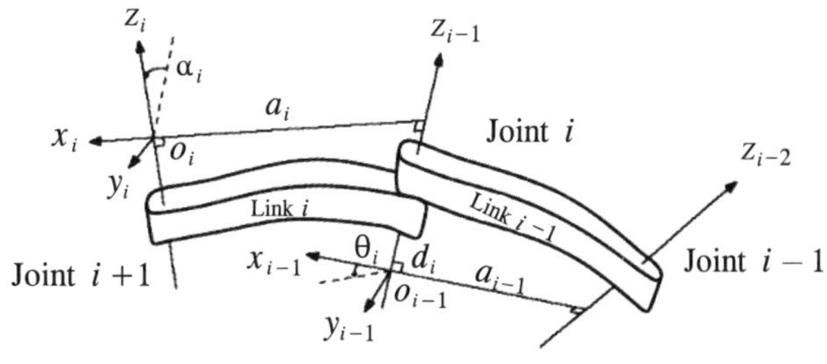

Fig.2-4 The Relationship Between Robot Links

图 2-4 机器人连杆间的关系

通过距离和方位轴，我们可得到不同的关节之间的关系，进而得到底座和末端执行器之间的关系。一旦我们有了这些变换，我们就可以计算关联部

We can go through the chain and then attach frames to the different joints and then find the transformation between the joints in order to find the relationship between the base frame and the end factor frame. Once we have those transformation, then we can compute the total transformation between successive things, and we can find the local transformation. Once we know the geometry. That is, we know where the end factor is and where each link is with respect to the others.

件之间的总变换，我们也可以找到局部变换。一旦我们知道了这层几何关系。我们也就知道了末端执行器的位置，以及每个连杆相对于其他连杆的位置。

微信扫一扫

教学视频

Lecture 2-3
Introduction to the DH Parameters
2-3 介绍 DH 参数

你可以在连杆的质心自由地放置坐标系。如果你想定义这些坐标系，我们就必须学习 DH 参数。我们来看一下 DH 参数，有四个参数，分别是连杆长度，连杆转角，连杆偏置和关节角度。如图 2-5 所示。

You can place the coordinate system freely at the center of mass of the link. If we want to define those frames, we must learn DH parameters. Let's look at DH parameters. Four DH parameters:[α_i, a_i, d_i, θ_i]: a_i: link length, α_i: link twist, d_i: link offset, θ_i: joint angle. As it is shown in Fig. 2-5.

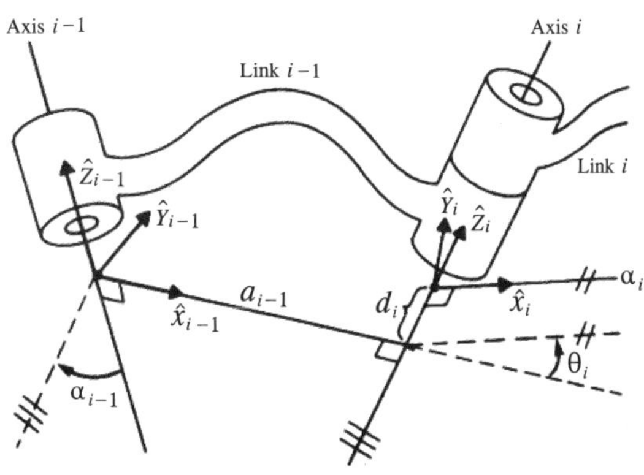

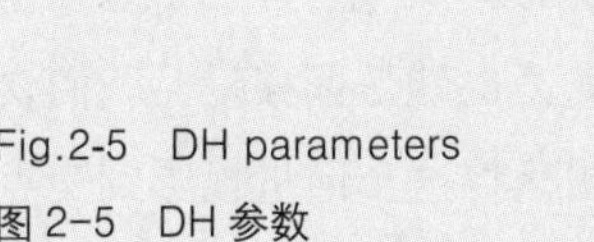
Fig.2-5 DH parameters
图 2-5 DH 参数

由于 DH 参数是一组只有一个变量的集合，其中三个对于每个连杆来说是不变的。d_i 对于移动关节是可变的，而θ_i 对于转动关节是可变的。

Since DH parameters is a sets of collection only one variable, three of these will be constant for each link. d_i will be variable for prismatic joints and θ_i will be variable for revolute joints.

轴 i-1 和轴 i 是通过连杆 i 连接起来的，如果我们取连杆

Axis i-1 and axis i are connected somehow through the link i. so if we take this link i, axis i, Axis i-1.So what are things that are constant? The distance.

We take the common perpendicular, and that will give us this distance. So we call it "a", so now you have to pay attention to the notation because we are going to describe link i-1 with the parameter a_{i-1}.

So if I slide axis i along this common perpendicular, and I come to the intersection, there will be an angle, a twist angle. We call it the link twist, and it is the parameter a_{i-1}, which measures this angle. And what we will do is we measure the angle along the vector Axis i minus one in the right hand sense. So a_{i-1} is the second parameter.

In fact, we are going to see that in total we need four parameters, one of them is variable, the joint angle or joint displacement. And now we identify two, they are all constant. If we move further, we are going to have another link. Now that other link will have another common normal. And this common normal will be between axis i and axis i+1. So that common normal will intersect with axis i. It will intersect at some location with axis i. And there is one more parameter, which is the distance , this offset. So what we do is we will project this vector on that intersection point, and then we can measure the distance d_i. And now you can see the angle. So di is defined by as the link offset, and this di is going to be constant for revolute joints.But for prismatic joints, it's really the direction along which the joint is going to affect the motion of the following link. So d_i will be variable if the joint is prismatic. So for a revolute joint, theta is the variable. So if like in this figure, this is a revolute joint as in here, this θ will be the variable. So θ_i is called the joint angle, and it's

i，轴 i，轴 i-1。那么什么是常数呢？距离。取公垂线，就得到这个距离。我们称它为 a，现在要注意这个符号，这里我们要用参数 a_{i-1} 来描述连杆 i-1。

如果我沿着公垂线滑动 i 轴，就会到达交点，就会有一个角，一个扭角。我们称它为连杆扭转，它是参数α_{i-1}，用来测量这个角度。测量这个角的方法是沿着向量轴 i-1 用右手法则来测量。α_{i-1} 是第二个参数。

实际上，我们总共需要四个参数，其中一个是可变的，关节角或关节位移。现在我们确定了两个，它们都是常数。如果我们继续，我们会有另一个连杆。现在另一个连杆就有了另一条公法线。公法线在 i 轴和 i+1 轴之间。所以公法线与 i 轴相交。它会在某个位置与 i 轴相交。还有一个参数，就是距离，也就是偏移量。我们要做的是将这个向量投影到交点上，然后我们就可以测量距离 d_i。现在你可以看到这个角度了。d_i 被定义为连杆的偏移量，这个 d_i 对于转动关节来说是常数。但对于移动关节，它就是关节影响下一个连杆运动的方向。所以如果关节是移动关节 d_i 是可变的。对于一个转动关节，θ_i 是变量。如果像这个图一样，这是一个转动关节，这个θ_i 就是变量。θ_i 被称为关节角,对于移动关节它是可变的。

旋转关节允许沿两个连杆之间的圆弧运动。棱柱(平动、线性、直线)关节允许沿两个连杆之间的直线运动。通过这种常规的方法获得了这四个参数，现在我们就基本上能够定义 DH 参数了。

variable for resolute joint.

Rotational joints allow motion along a circular arc between two links. Prismatic (Translational, Linear, Rectilinear) joints allow motion along a straight line between two links. So with this convention and those four parameters, now we can basically defined the DH parameters.

微信扫一扫

教学视频

Lecture 2-4
Introduction to the Frame Assignment(1)
2-4 坐标系介绍（1）

What is a frame? What is the most critical thing about the frame? First, we really need to decide a rule about how we selects the origin of those frames. And we also need to decide the axes. A frame is the origin and the axes, so you have X, Y and Z. And have to make some decision also about the axes.

We selected i-1 and i, and we said these are the axes, and we know the distance between them, and the offset between them. So these axes have a very important role. You want your frame to be aligned with those axes. So you want these joint axes to be the axes that define one of the vectors, basically X, Y and Z.

This origin has to again make use of the information of DH parameters. We said we are using the common normal. This point of intersection with the common normal is very important. So now we are going to take the origin of our frames at the intersection with the common normal.

So we know the origin, and we are going to say X-axis has to be orthogonal to Z. This axis A-1 is orthogonal Z-1 because it is the axis orthogonal to the joint anyway. So we take, X-axis in this direction. The Y is defined in the right hand rule.

坐标系是什么？建立坐标系最重要的是什么？首先，我们需要决定一个关于如何选择坐标系的原点规则。我们还需要确定坐标轴。一个坐标系是原点和坐标轴组成的，所以你有 X，Y 和 Z 轴，还需要确定坐标轴。

我们选择 i - 1 和 i 关节轴，我们还知道它们之间的距离，它们之间的偏移。所以这些轴非常重要。你希望你的坐标系与这些轴对齐。你想要这些关节轴是定义其中一个向量的坐标轴，是 X，Y 还是 Z 轴。

原点的确定必须再次利用 DH 参数的信息。我们说过要用公法线。这个点与公法线的交点非常重要。现在我们要取坐标系的原点在与公法线的交点处。

我们确定了原点，我们知道 X 轴与 Z 轴正交，这个轴 A - 1 与 z_i-1 正交因为它是与关节正交的轴。我们取这个方向为 X 轴方向。Y 轴通过右手定则定义。

这仅仅是对建立坐标系方法的简单介绍。事实上，它要复杂得多，因为有很多特殊情况，可能有相交轴，平行轴，还有最后一个末端坐标系仍是没有定义的，你必须定义末端坐标系，方法也很多。这就是为什么我们总是尽量减少参数，得到零参数，这会使问题变简单。

This is simple introduction of the frame assignment. In reality, it is much more complicated because there are all these special cases where you have intersecting axes, parallel axes, last frame that is still not defined, so you have to define last frome by using different ways. And that's why the convention of always reducing, getting zero parameters, and this will make the problem easier.

微信扫一扫

教学视频

Lecture 2-5
Introduction to the Frame Assignment(2)
2–5 工业机器人坐标系（2）

You have the axes. Take the intersection, and pick the origins. Along the origins, you place your Z-axis. And then you defined the X-axis along the common normals. And that's it. So you have four steps: finding the common normal, finding the origin of their intersection, taking the Z axis along the joint axis from the point of the the origins you selected, and then placing the X axis.

Let's see the cause of intersecting joint axes, which happens very often in the beginning and the end of the mechanism. In the beginning, usually joint 1 and 2 are intersecting, and joint 4, 5, and 6 are intersecting. In six degree of freedom manipulators commonly have three intersecting joints.The mechanics of that is well understood and it's quite easy to build. And also, there are a lot of advantages in terms of the workspace and the freedom, except they have problems with respect to singularities. So we know this point, and what we said is that we picked the intersection as the origin. And now this is the origin for frame I, and we place the Z-axis, Z+1 axis, that the question is where do you select your X direction. If you place your x_i in this direction,your twist angle

现在你有一个坐标轴，找到交点，然后选择原点。沿着原点，放置 Z 轴。然后沿着公法线定义 X 轴。就是这样。可分为四个步骤：找到公法线，找到它们的交点，沿着关节轴取 z 轴，然后放置 x 轴。

我们来看看关节轴相交的原因，这种情况在机构的开始和结束时经常发生。通常，关节 1 和 2 是相交的，关节 4、5、6 也是相交的。对于六自由度的机器人，一般带有三个交叉关节的手腕。它的机制很好理解，也很容易构建。同时，这种结构在工作空间和自由度方面也有很多优势，除了奇点方面的问题外。我们可以选择交点作为原点。现在这是坐标系 I 的原点，我们放置 Z 轴，Z+1 轴，问题是你要怎么选择 X 轴的方向。如果你把 x_i 放在这个方向，你的扭转角就会这样测量。这样的话，扭转角是这个方向，如果你把它放在相反的方向，则是另一个方向。如果你把 X 轴放在不同方向，你会得到扭转角的不同定义。一旦你有了 X 轴位置，你就定义了扭转角。扭转角的方向和符号取决于你对 X 轴的选择。

下面举例说一下第一个连杆的定义方法。对于转动关节，我们要做的是，当θ等于 0 时，固定的坐标系和运动坐标系是相同的。我们知道，固定的框架依附于固定底座，这可自由选择。由于 A，α都等于 0，这就获得了最简单的坐标系定义形式。唯一的变量是θ，当θ为 0 时，两个坐标系重合。所以当你旋转这个关节时，θ就等于你要旋转的角度。对于移动关节，一开始，也是让这两个坐标系重合。当变量 d_i 为 0 时，这两个坐标系重合。

对于最后一个连杆，如果是一个转动关节，末端坐标系就选择 d_n 等于 0 的坐标系，这取决于后面的框架。也就是说，$d_n=0$，这个坐标系只有一个关节转角变量，当转角$\theta_n=0$ 时，坐标系 x_n 和坐标系 x_n-1 对齐。旋转的时候，就可以测量转角的角度了。对于移动关节，同样把转角设为 0。当 $d_n=0$ 时，坐标系 x_n 与 x_n-1 对齐。这些是

will be measured about that. So it will be in this sense in this direction, and if you place it in the opposite direction so this is one direction. Once you have place xi, you have defined your alpha. So this direction and the sign of alpha depends the on your picking of X.

So here is an example of the defination method of the first link. I'm taking a revolute link. What we would like to do for a revolute link is to almost say that the fixed frame and the moving frame are identical when θ is equal to zero. As we known, the fixed frame attached fixed base, it all comes to freedom of choice. So you are really setting A and α to be 0, and d_i equal to zero for the zero for the revolute joint and the zero of theta is when the two frames are identical.

For the last link, if it's a revolute joint, we are going to select the frame that dn equal to zero, which depends on the following frame. That is we are saying d_n equal to zero, and that frame just measures the angle θ_n. and when θ_n equals to zero, basically we have x_n, and x_n-1 and x_n align. θ measures this angle between the x-axes basically, as you rotate. And for the prismatic joint, we do the same thing to set θ_n to zero. That is we have when d_n equal to zero, x_n comes down to be aligned with x_n-1. so these are the conventions that you are going to try to enforce in your frame assignment. And using these, you will end up with the foundation for the forword kinematics.

设置坐标系的规则。利用这些，你将得到正向运动学的基础能力。

微信扫一扫

教学视频

Lecture 2-6
Introduction to the Frame Assignment(3)

2-6 工业机器人坐标系（3）

让我们回顾一下关于 DH 参数和坐标系的内容。我们需要为每个关节引入这四个参数，a_i，α_i，θ_i，d_i。a_i 用于测量 z_i 和 z_i+1 沿 x_i 轴的距离。α_i 测量 z_i 轴和 z_i+1 关于 x_i 轴的夹角。d_i 表示 x_i-1 和 x_i 沿着 z_i 轴的距离。θ_i 表示 x 轴绕 z_i 的夹角。

举个例子，一个平面机器人，这个平面机器人只有三个转动关节。三个转角θ_1、θ_2 和 θ_3。关节轴垂直于平面。那么我们如何选择原点呢？为了选择原点，你可以看到三个平行轴，你需要公法线。平行轴之间有很多可能的公法线，但是因为这个在平面上，我们要用这个平面上的公法线。如果公法线相交于这一点，这一点即为要分配的坐标系的原点。

对于坐标系 1，公法线在这里相交，z1 朝向平面外，X1 沿着公法线，再确定了 y1，就

Let's start by reviewing what we did about the DH parameters and the frame attachment. So what we said is we need to introduce four parameters for each joint, a_i, α_i, θ_i, and d_i. a_i is measuring the distance between z_i and z_i+1 along the x_i axis. Alpha I measures the angle between axis z_i and z_i+1 about the x_i axis in the right hand sense. Di measures the distance between the x-axis, x_i-1, and x_i along the z_i axis. And theta I measures the angle between the x_i-1 and x_i about z_i.

let's take an example. I'm going to take a planar robot, and this planar robot is just a set of three revolute joints. So we are talking about θ_1, θ_2, and θ_3. The joint axes are coming perpendicular to the plane. So how do we pick the origins? In order to pick the origins so you see three parallel axes. You need the common normal. Between parallel axes may have many possible common normal.but because this in the plane, we are going to use the common normal in that plane. The common normal are intersecting at this point, this point will become the origin of these frames that we are going to assign.

So for frame1, the common normal is intersecting here, the z1 will be out of the page. X1 is along the common normal. And y1 complete the direct frame, so basically you have this as the first frame. X2 is along the common normal, and z2 is coming out of the plane, y2 is here. And we are placing the last frame. We have the z3, and we are placing the origin, so x3 is taking along the direction perpendicular to the z3 and along the direction to L3, and that measures the angle θ_3. So between these frames, the only variables that you are going to see is θ_1, θ_2, θ_3, and now you need to introduce the first frame. So for the first frame,we said we are going to simplify. We are not going to select a z0 in an arbitrary direction. We select z0 along z1. so this way, we will select the x0 to be coincident with x1 when θ_1 is equal to 0. And θ_1 is measured from here, so x0 will be along this direction.

完成了坐标系，这个作为第一个坐标系。X2 沿着公法线，z2 朝向平面外，y2 在这里。然后我们放置最后一个坐标系。我们有了 z3，我们再放置原点，如果 x3 垂直于 z3 的方向并且沿着 L3 的方向，就可以轻易得到旋转角θ_3。在这些坐标系中，你唯一能看到的变量是θ_1、θ_2、θ_3，现在你需要引入 0 坐标系。对于 0 坐标系，我们说过要尽量简化。我们不会选择任意方向的 z0，要沿着 z1 选择 z0。这样，当$\theta_1=0$时，坐标系 x0 就与坐标系 x1 重合。θ_1是可以轻易测量出来的，所以 x0 是沿着这个方向的。

With this frame assignment,we can defined for each of the joint those parameters. And we say joint 1. For joint 1, α_0 is equal to 0. There is no angle between z0 and z1. No distance between the two. And the only variable is θ_1. So we are going to go through this one by one, and because of the fact that in this case the only variable that are going to be Extra-introduced are due to L1, L2. That means A1 is going to be L1, and A2 is going to be L2.

有了坐标系，我们可以为每个关节定义 DH 参数。首先来谈关节 1。对于关节 1，$\alpha_0=0$。z0 和 z1 之间没有夹角。两者之间也没有距离。唯一的变量是θ_1。我们将一个接一个地推导，因为在这种情况下唯一额外引入的变量是 L1 和 L2。这意味着 A1 等于 L1，A2 等于 L2。

Usually, we add one more column where we say configurations shown. In this shown, you set the value of the variable. With those four parameters and this table, we should be able to describe the forward kinematics. That is we should be able to describe the position and orientation of the end- effector.

通常，我们再加一列表示构型的列。在这一列中，你可以设置变量的值。现在有了这四个参数和这个表格，我们就能描述正向运动学了。也就是说我们能够描述末端执行器的位置和方向了。

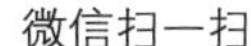

Lecture 3-1
Manual Operation of Industrial Robots (1)

教学视频

3-1 工业机器人的手工操作（1）

工业机器人示教器（Flexpendant）是一种手持式操作装置，由硬件和软件组成，用于执行与操作和工业机器人系统有关的许多任务：运行程序、参数配置、修改机器人程序等。Flexpendant可在恶劣的工作环境下持续工作，其触摸屏易于清洁，且防水、防油、防溅泼。Flexpendant本身就是一台完整的计算机，通过集成线缆和接头连接到控制器。

1. 工业机器人开关机

机器人实际操作的第一步就是开机，将机器人控制柜上的总电源旋钮从【OFF】扭转到【ON】即可，如图 3-1 所示。关机时，将电源开关从【ON】扭转到【OFF】即可。

2. 示教器

在示教器上，绝大多数的操作都是在触摸屏上完成的，同时也保留了必要的按钮和操作装置，如图 3-1 所示。

Flexpendant is a hand-held operating device composed of hardware and software, used to perform and operate many tasks related to industrial robot systems: running programs, parameter configuration, and modifying robot programs. Flexpendant can work continuously in harsh working environment, its touch screen is easy to clean, and it is waterproof, oil-proof and splash-proof. The Flexpendant itself is a complete computer, connected to the controller via integrated cables and connectors.

1. Startup and Shundown of Industrial Robot

The first step in the actual operation of the robot is to turn it on. Turn the main power knob on the robot control cabinet from [OFF] to [ON], as shown in Fig. 3-1. When shutting down, turn the power knob from [ON] to [OFF].

2. Flexpendant

On the Flexpendant, most of the operations are done on the touch screen, and the necessary buttons and operating devices are also reserved, as shown in Fig. 3-1.

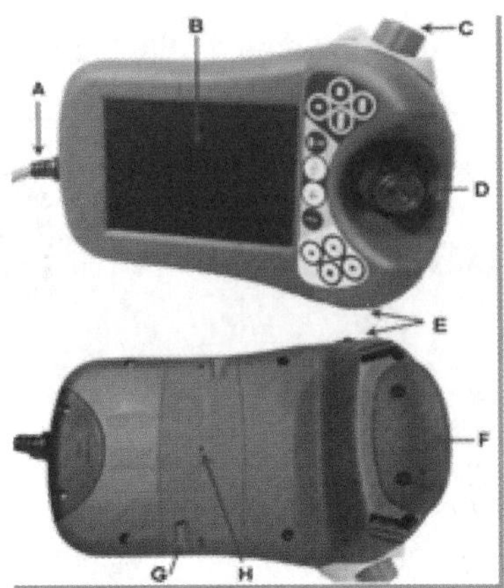

Fig.3-1 Description for the Structure of the Flexpendant
A 链接电缆 B 触摸屏
C 急停开关 D 手动操纵摇杆
E USB 端口 F 使能器按钮
G 触摸屏用笔 H 示教器复位按钮
图 3-1 示教器的结构说明

A Link cable B Touch screen C Emergency stop switch D Manual joystick E USB port F Enabler button G Pen for touch screen H Reset button for Flexpendant

The enabler button of the industrial robot is set to ensure the personal safety of the operator. Only when the enabler button is pressed and the motor is turned on can the robot be manually operated and debugged. When there is a danger, people will instinctively loosen or press the enabler button, and the robot will immediately stop to ensure safety. Hand-held Flexpendant as shown in Fig. 3-2 below.

使能器按钮是工业机器人为保证操作人员人身安全而设置的。只有在按下使能器按钮，并保持在电机开启的状态，才可对机器人进行手动的操作与程序的调试。当发生危险时，人会本能地将使能器按钮松开或按紧，则机器人会马上停下来，保证安全。手持示教器如图 3-2 所示。

Fig.3-2 Correct Use of Hand-held Flexpendant
图 3-2 正确手持示教器

3.设定示教器的语言设置

示教器出厂时，默认的显示语言是英语，为了方便操作，下面介绍把显示语言设定为中文的操作步骤。

3. Language Setting for Flexpendant

When the Flexpendant is shipped from the factory, the default display language is English. In order to facilitate the operation, the following describes the operation steps for setting the display language to Chinese.

(1)单击左上角主菜单按钮。

(2)选择“Control Panel”。

如图 3-3 所示。

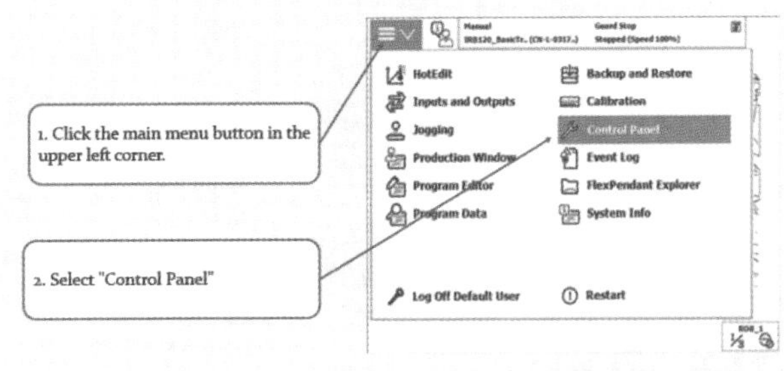

Fig.3-3 Operation Step 1 and 2

图 3-3 操作步骤 1 和 2

(3) 选择 “Language”。

如图 3-4 所示。

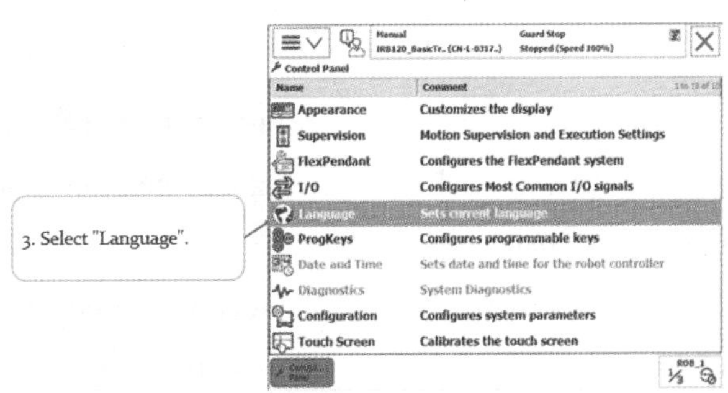

Fig.3-4 Operation Step 3

图 3-4 操作步骤 3

(4) 选择 “Chinese”。

(5) 单击 “OK”。

如图 3-5 所示。

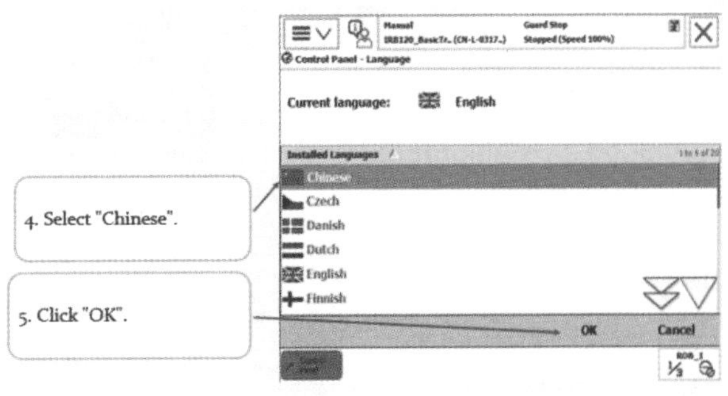

Fig.3-5 Operation Step 4 and 5

图 3-5 操作步骤 4 和 5

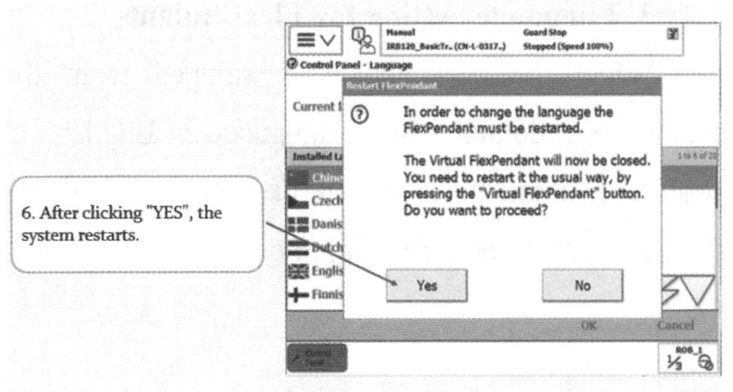

（6）点击“YES”后，系统重启。

如图 3-6 所示。

Fig.3-6 Operation Step 6

图 3-6 操作步骤 6

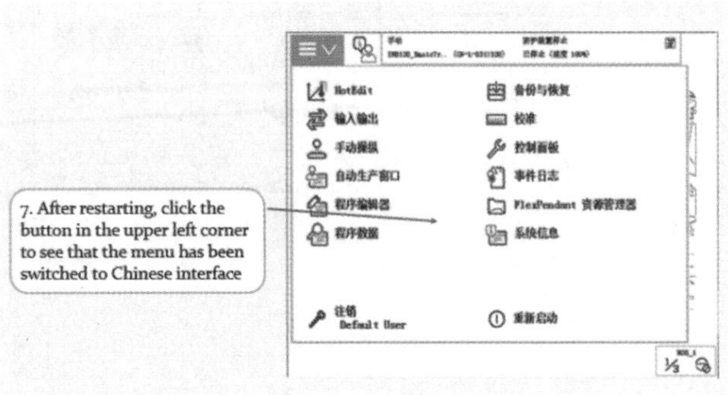

（7）重启后，单击左上角按钮就能看到菜单已切换成中文界面。

如图 3-7 所示。

Fig.3-7 Operation Step 7

图 3-7 操作步骤 7

4.Time Setting for the Industrial Robot System

In order to facilitate file management and fault review and management, the time of the robot system should be set to the time of the local time zone before various operations. The steps are as follows.

4. 设定工业机器人系统的时间

为了方便进行文件的管理和故障的查阅与管理，在进行各种操作之前要将机器人系统的时间设定为本地时区的时间。步骤如下。

（1）单击左上角主菜单按钮。

（2）选择“控制面板”。

如图 3-8 所示。

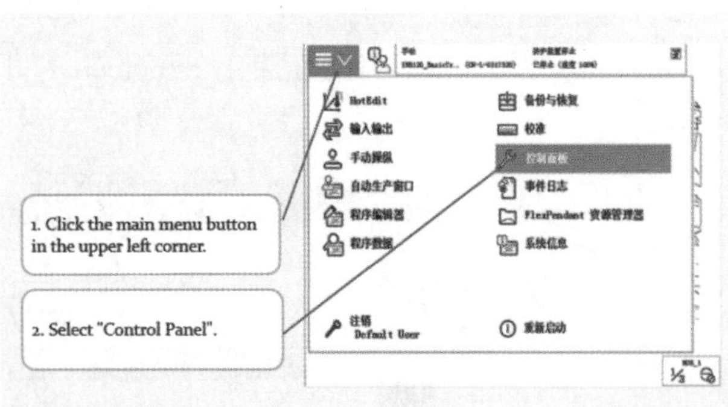

Fig.3-8 Time Setting Step 1

图 3-8 时间设置步骤 1

（3）选择“日期和时间”。

如图 3-9 所示。

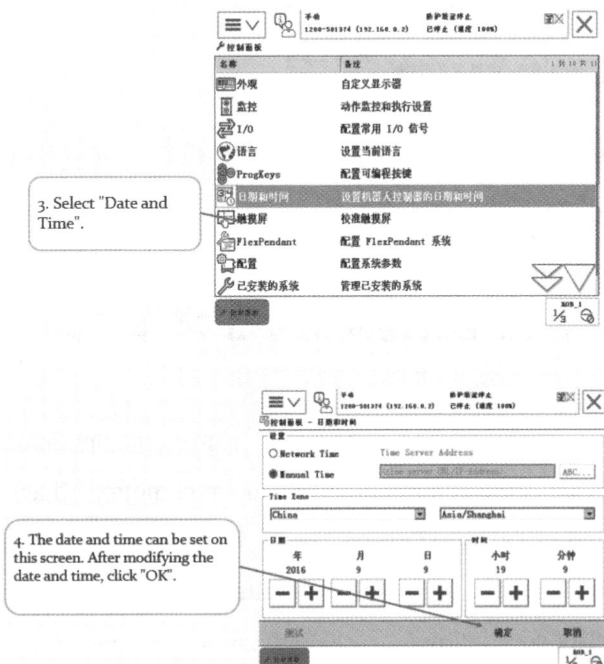

Fig.3-9 Time Setting Step 2

图 3-9 时间设置步骤 2

（4）在此画面就能对日期和时间进行设定。日期和时间修改完成后，单击“确定”。

如图 3-10 所示。

Fig.3-10 Time Setting Step 3

图 3-10 时间设置步骤 3

5. 示教器事件日志查看

可以通过示教器画面上的状态栏进行ABB机器人常用信息及事件日志的查看：

A 机器人的状态（手动、全速手动和自动）。

B 机器人的系统信息。

C 机器人的电机状态。

D 机器人的程序运行状态。

E 当前机器人或外轴的使用状态。

单击画面中的状态栏就可以查看机器人的事件日志。

如图 3-11 所示。

5. View of the Event Log of the Flexpendant

You can view the common information and event logs of ABB robots through the status column on the Flexpendant screen:

A The status of the robot (manual, full-speed manual and automatic).

B The system information of the robot.

C The motor status of the robot.

D The program running status of the robot.

E The current status of the robot or external axis.

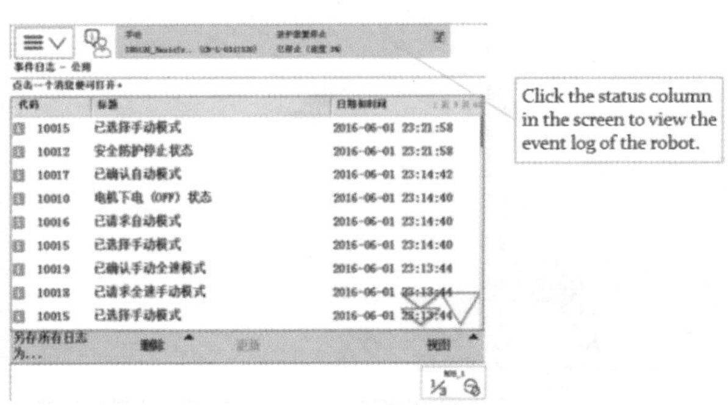

Fig.3-11 View of the Event Log

图 3-11 日志查看

Lecture 3-2
Manual Operation of Industrial Robots (2)

微信扫一扫

教学视频

3-2 工业机器人的手工操作（2）

There are three modes of manual manipulation of industrial robot motion: single-axis motion, linear motion and repositioning motion. The manual operation and incremental mode of single-axis motion are introduced below.

1. Single Axis Motion

Generally, an ABB robot uses six servo motors to drive the six joint axes of the robot, as shown in Fig. 3-12 below. Then every time the motion of a joint axis is manually manipulated, it is called a single-axis motion. The following is how to manually manipulate single-axis motion.

手动操纵工业机器人运动一共有三种模式：单轴运动、线性运动和重定位运动。下面介绍单轴运动手动操纵以及增量模式。

1. 单轴运动

一般地，ABB 机器人是由六个伺服电机分别驱动机器人的六个关节轴，如下图 3-12 所示。那么每次手动操纵一个关节轴的运动，就称之为单轴运动。以下就是手动操纵单轴运动的方法。

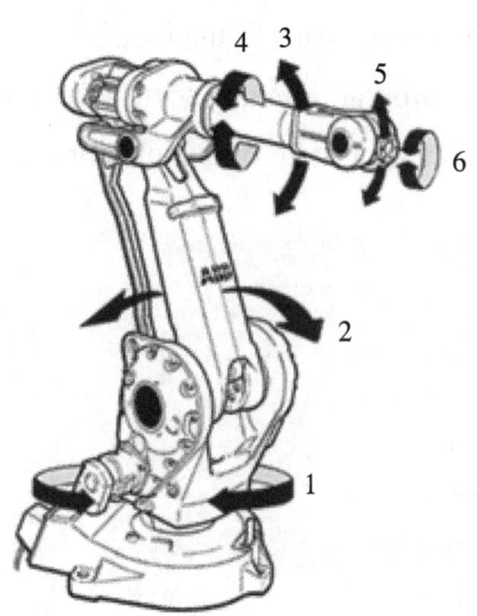

Fig.3-12 Six-Joint Axes of Industrial Robot

图 3-12 工业机器人六关节轴

（1）将控制柜上机器人状态钥匙切换到手动限速状态（小手标志）。

如图 3-13 所示。

电源总开关

急停开关

电机通电/复位

机器人状态

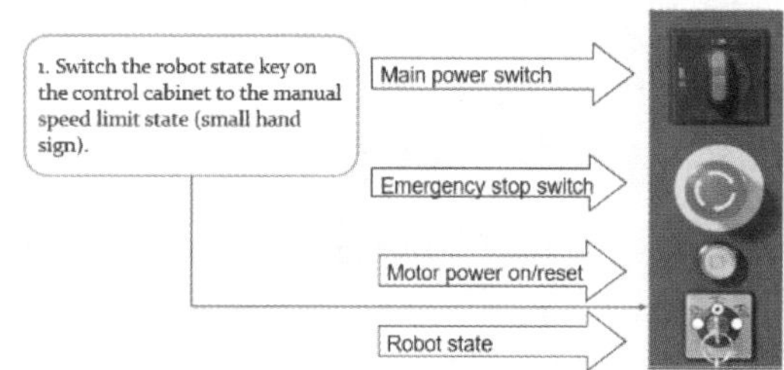

Fig.3-13　Switch the Robot State Step 1

图 3-13　切换机器人状态步骤 1

（2）在状态栏中，确认机器人的状态已切换为“手动”。

（3）单击左上角主菜单按钮。

如图 3-14 所示。

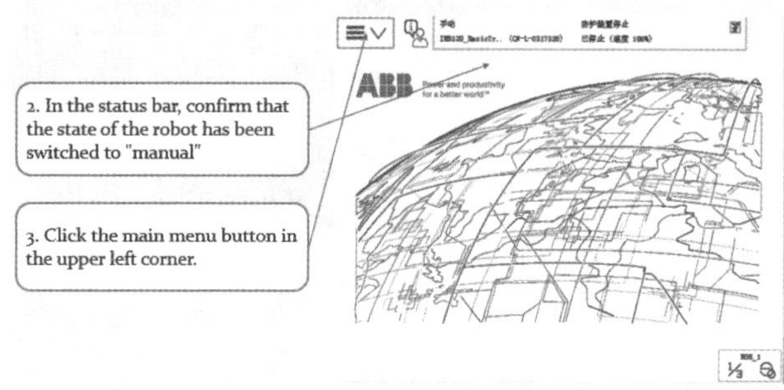

Fig.3-14　Operation Step 2 and 3

图 3-14　操作步骤 2 和 3

（4）选择“手动操纵”。

如图 3-15 所示。

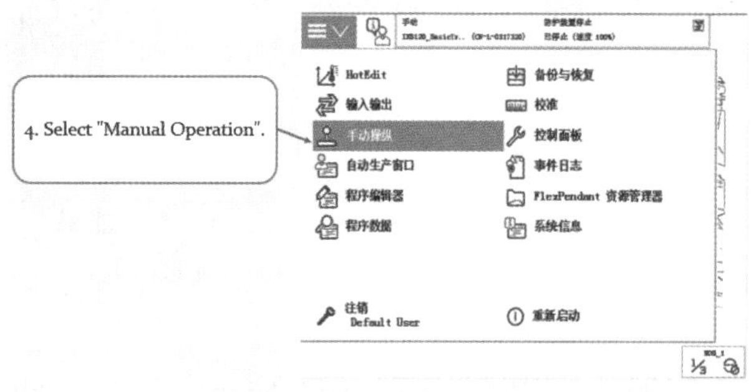

Fig.3-15　Operation Step 4

图 3-15　操作步骤 4

（5）单击“动作模式”。

如图 3-16 所示。

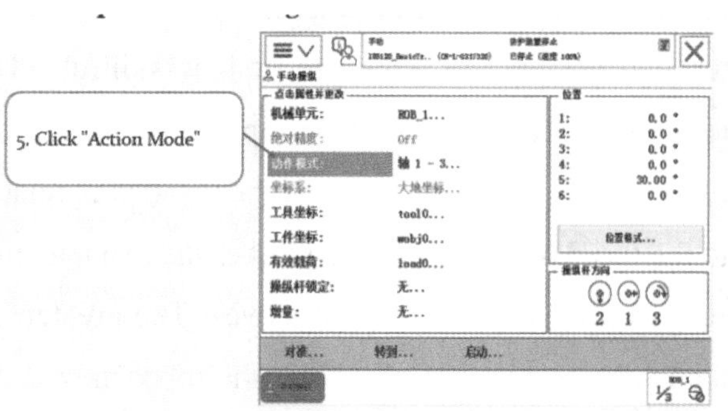

Fig.3-16　Operation Step 5

图 3-16　操作步骤 5

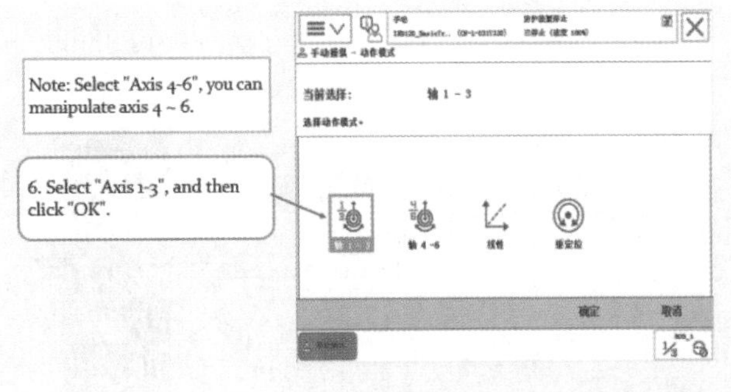

说明：选中“轴 4-6”，就可以操纵轴 4～6。

（6）选中“轴 1-3”，然后单击“确定”。

如图 3-17 所示。

Fig.3-17 Operation Step 6

图 3-17 操作步骤 6

7. Press the enabler button with your left hand to enter the "motor on" state

（7）用左手按下使能按钮，进入“电机开启”状态。

如图 3-18 所示。

Fig.3-18 Operation Step 7

图 3-18 操作步骤 7

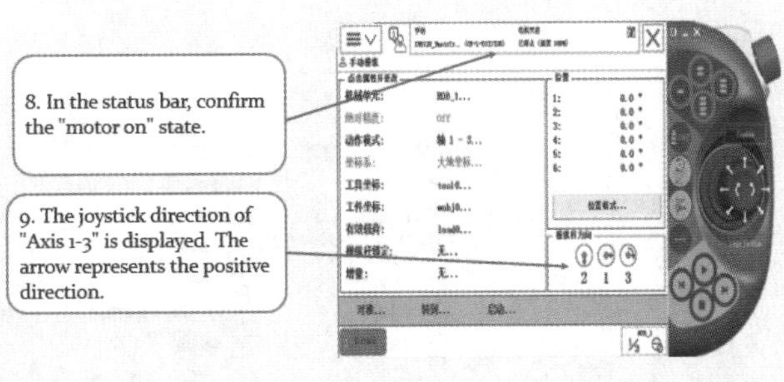

（8）在状态栏中，确认“电机开启”状态。

（9）显示“轴 1-3”的操纵杆方向。箭头代表正方向。

如图 3-19 所示。

Fig.3-19 Operation Step 8 and 9

图 3-19 操作步骤 8 和 9

The 4-6 axes can be manipulated by the same method as above. When manipulating industrial robots through the joystick, the use of it is skillful. The joystick of the robot can be compared to the throttle of a car. The operating range of the joystick is related to the speed of the robot. The smaller the manipulation range is, the slower the robot moves. The greater the manipulation range is, the faster the robot moves. So

通过上面同样的方法可以操纵 4-6 轴，在通过操纵杆操纵工业机器人时，操纵杆的使用是有技巧的，可以将机器人的操纵杆比作汽车的节气门，操纵杆的操纵幅度是与机器人的运动速度相关的。操纵幅度较小，则机器人运动速度较慢。

操纵幅度较大，则机器人运动速度较快。所以在操作时，尽量以小幅度操纵使机器人慢慢运动来开始手动操纵学习。

2. 增量运动

如果对使用操纵杆通过位移幅度来控制机器人运动的速度不熟练，可以使用“增量”模式来控制机器人的运动。

在增量模式下，操纵杆每位移一次，机器人就移动一步。如果操纵杆持续一秒或数秒钟，机器人就会持续移动（速率为 10 步/s）。增量的移动距离及弧度如下表 3-1 所示。

表 3-1 增量移动距离及弧度

增量	移动距离/mm	弧度/rad
小	0.05	0.0005
中	1	0.004
大	5	0.009
用户	自定义	自定义

增量运动操作步骤如下：

（1）选中“增量”。

如图 3-20 所示。

when you operate, try to make the robot move slowly with a small range of manipulation to start manual manipulation learning.

2. Incremental Motion

If you are not familiar with using the joystick to control the speed of the robot motion through the displacement amplitude. Then you can use the “incremental” mode to control the robot’s motion.

In the incremental mode, each time the joystick moves, the robot moves one step. If the joystick lasts for one or several seconds, the robot will continue to move (at a rate of 10 steps/s). The incremental motion distance and radians are shown in Table 3-1 below.

Table 3-1 Incremental Motion Distance and Radians

Increment	Motion distance/mm	Radians/rad
Small	0.05	0.0005
Medium	1	0.004
Large	5	0.009
User	customized	customized

The steps of incremental motion are as follows:

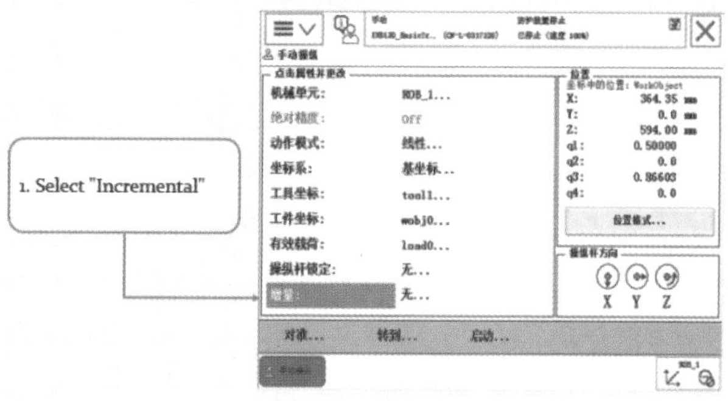

Fig.3-20 Incremental Motion Operation Step 1

图 3-20 增量运动操作步骤 1

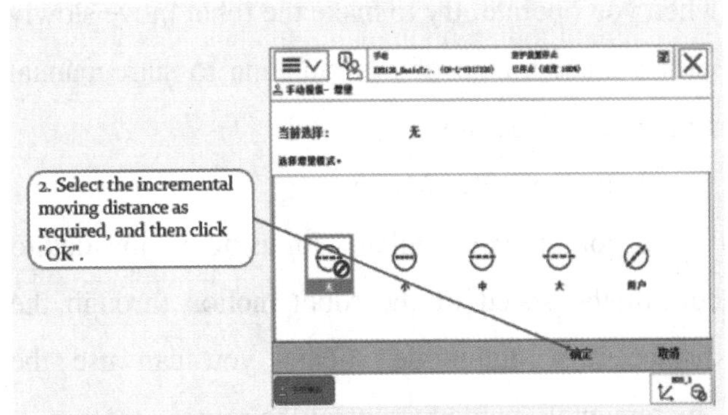

（2）根据需要选择增量的移动距离，然后单击“确定”。如图 3-21 所示。

Fig.3-21 Incremental Motion Operation Step 2

图 3-21 增量运动操作步骤 2

Lecture 3-3

Manual Operation of Industrial Robots (3)

微信扫一扫

教学视频

3-3 工业机器人的手工操作（3）

1. 线性运动

机器人的线性运动是指安装在机器人第六轴法兰盘上工具的 TCP 在空间中做线性运动。以下是手动操纵线性运动的方法。

（1）选择“手动操纵”。如图 3-22 所示。

1. Linear Motion

The linear motion of the robot refers to the linear motion of the TCP of the tool installed on the sixth axis flange of the robot in space. The method of manually manipulating linear motion is as follows.

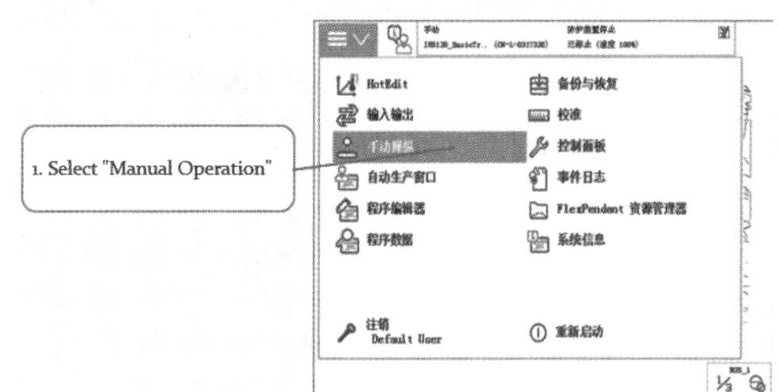

Fig.3-22　Operation Step 1

图 3-22　操作步骤 1

（2）单击“动作模式”。如图 3-23 所示。

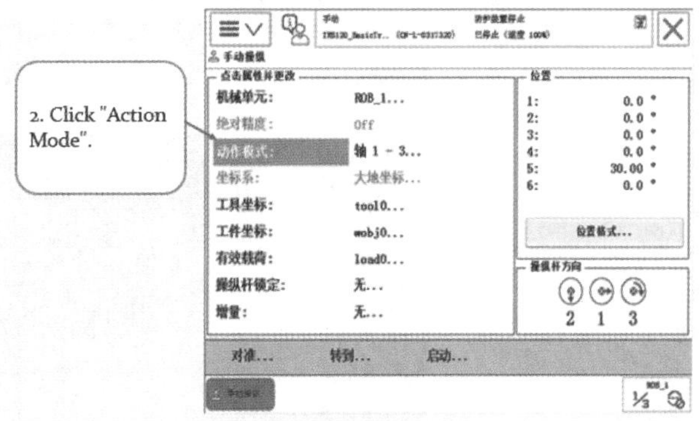

Fig.3-23　Operation Step 2

图 3-23　操作步骤 2

（3）选择“线性”，然后单击“确定”。

如图 3-24 所示。

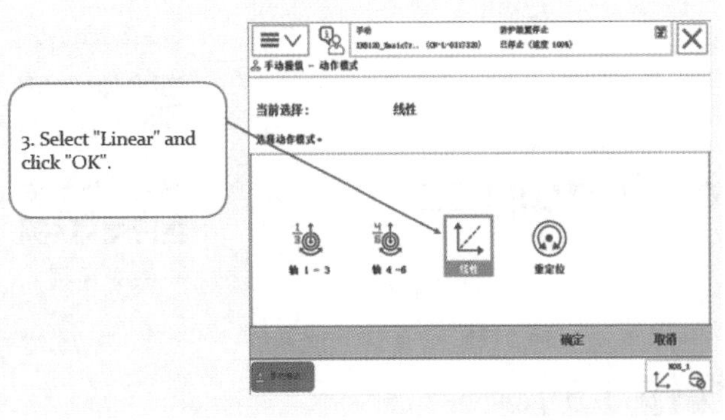

Fig.3-24 Operation Step 3

图 3-24 操作步骤 3

（4）单击“工具坐标”。

如图 3-25 所示。

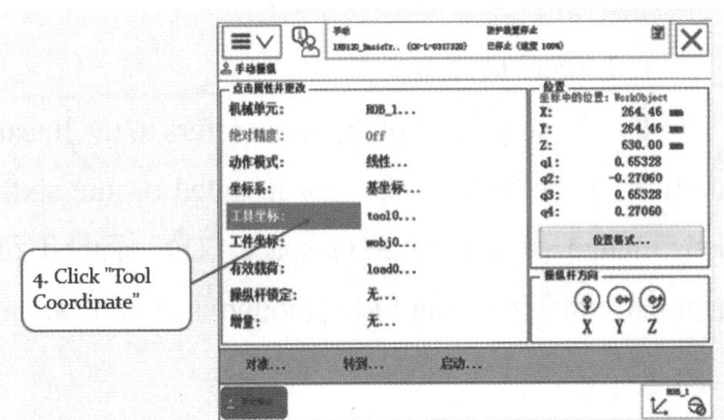

Fig.3-25 Operation Step 4

图 3-25 操作步骤 4

The corresponding tool in the “Tool Coordinate” should be specified for the linear motion of the robot.

机器人的线性运动要在“工具坐标”中指定对应的工具。

（5）选择对应的工具“tool1”，然后单击“确定”。

如图 3-26 所示。

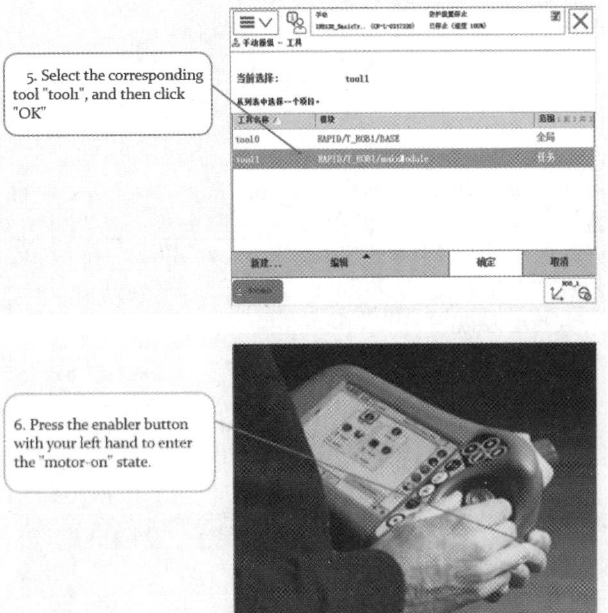

Fig.3-26 Operation Step 5

图 3-26 操作步骤 5

（6）用左手按下使能按钮，进入“电机开启”状态。

如图 3-27 所示。

Fig.3-27 Operation Step 6

图 3-27 操作步骤 6

（7）在状态中，确认“电机开启”状态。

（8）显示轴 X、Y、Z 的操纵杆方向。箭头代表正方向。

如图 3-28 所示。

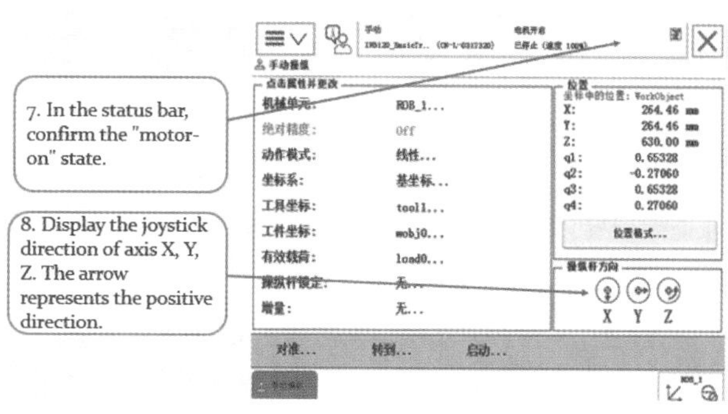

Fig.3-28　Operation Step 7 and 8

图 3-28　操作步骤 7 和 8

（9）操作示教器上的操纵杆，工具的 TCP 点在空间中做线性运动。

如图 3-29 所示。

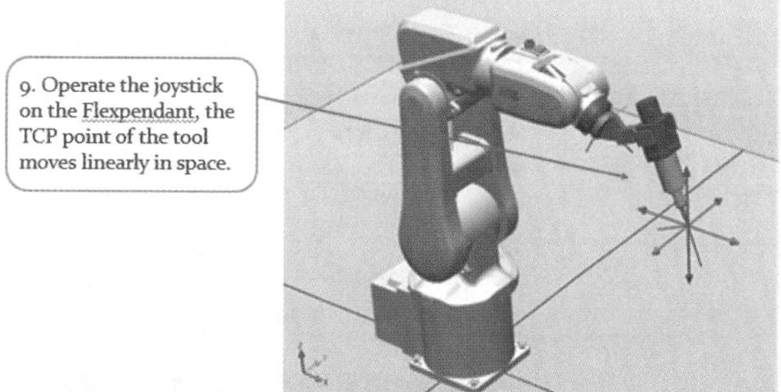

Fig.3-29　Operation Step 9

图 3-29　操作步骤 9

2. 重定位运动

机器人的重定位运动是指机器人第六轴法兰盘上的工具 TCP 点在空间中绕着坐标轴旋转的运动，也可以理解为机器人绕着工具 TCP 点做姿态调整的运动。以下就是手动操纵重定位运动的方法。

（1）选择“手动操纵”。

如图 3-30 所示。

2. Reposition Motion

The reposition motion of the robot refers to the movement of the tool TCP point on the sixth axis flange of the robot rotating around the coordinate axis in space, and can also be understood as the movement of the robot to adjust the posture around the tool TCP point. The method of manually manipulating reposition motion is as follows.

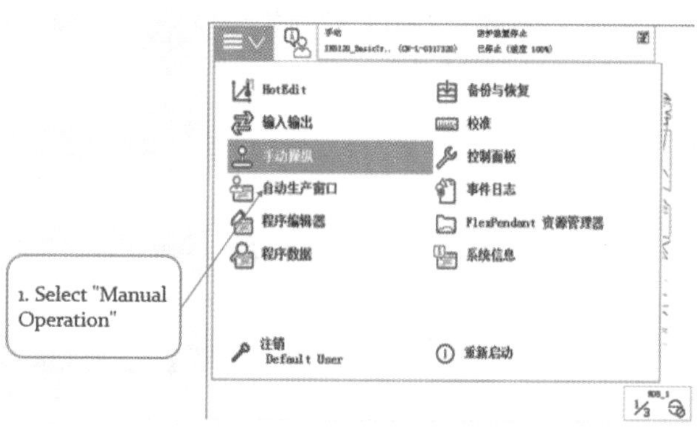

Fig.3-30　Operation Step 1

图 3-30　操作步骤 1

（2）单击“动作模式”。

如图 3-31 所示。

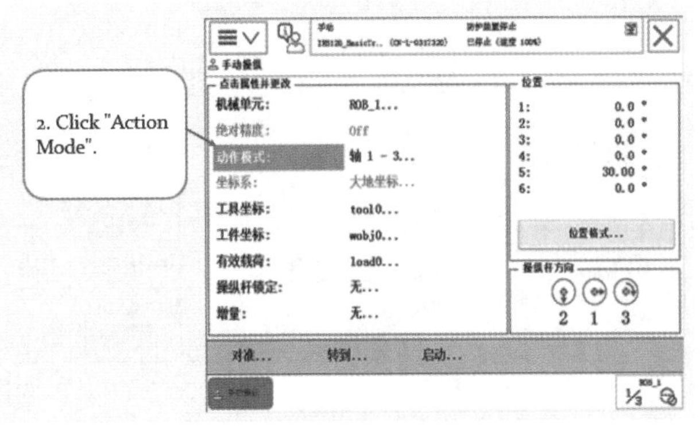

Fig.3-31 Operation Step 2

图 3-31 操作步骤 2

（3）选择“重定位”，然后单击“确定”。

如图 3-32 所示。

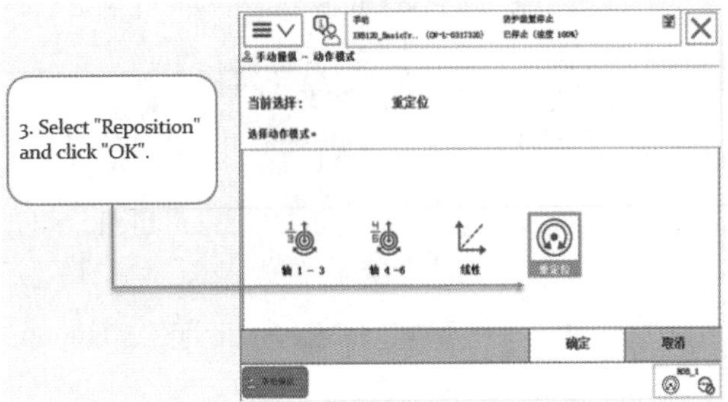

Fig.3-32 Operation Step 3

图 3-32 操作步骤 3

（4）单击“坐标系”。

如图 3-33 所示。

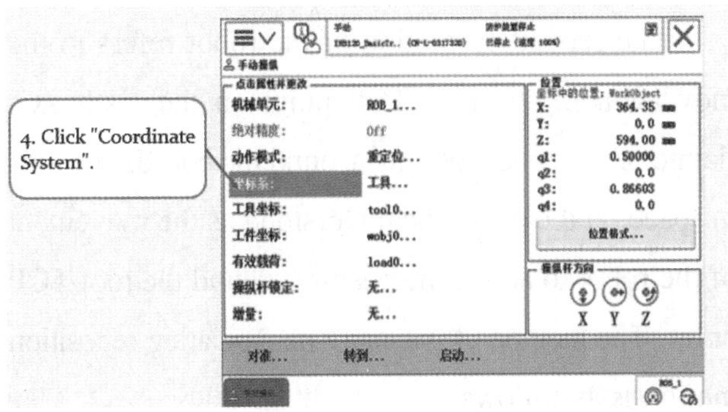

Fig.3-33 Operation Step 4

图 3-33 操作步骤 4

（5）选择“工具”，然后单击“确定”。

如图 3-34 所示。

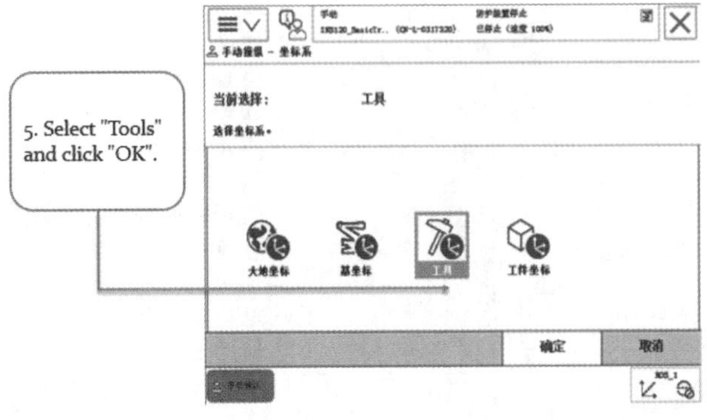

Fig.3-34 Operation Step 5

图 3-34 操作步骤 5

（6）单击“工具坐标”。

如图 3-35 所示。

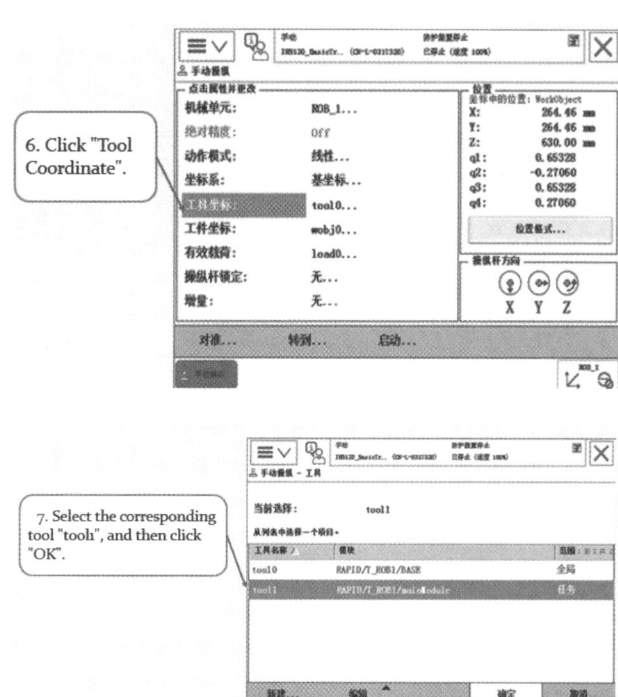

Fig.3-35 Operation Step 6

图 3-35 操作步骤 6

（7）选择对应的工具“tool1”，然后单击“确定”。

如图 3-36 所示。

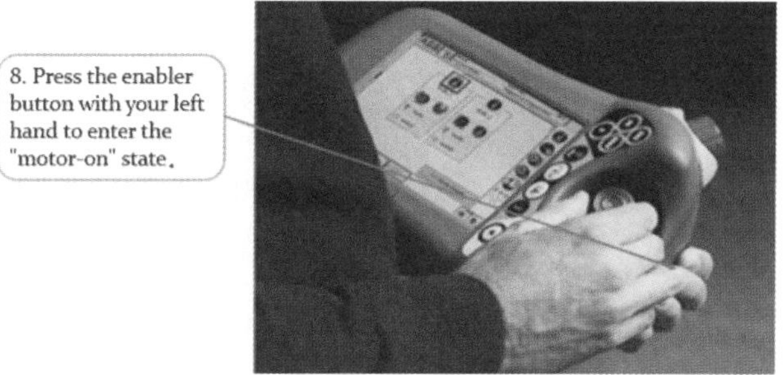

Fig.3-36 Operation Step 7

图 3-36 操作步骤 7

（8）用左手按下使能按钮，进入“电机开启”状态。

如图 3-37 所示。

Fig.3-37 Operation Step 8

图 3-37 操作步骤 8

（9）在状态中，确认“电机开启”状态。

（10）显示轴 X、Y、Z 的操纵杆方向。箭头代表正方向。

如图 3-38 所示。

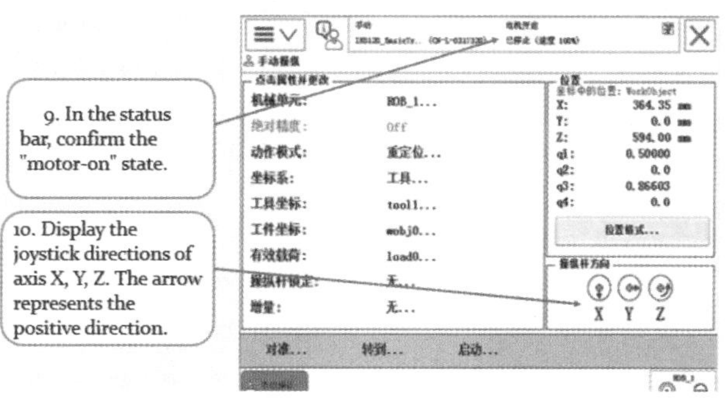

Fig.3-38 Operation Step 9 and 10

图 3-38 操作步骤 9 和 10

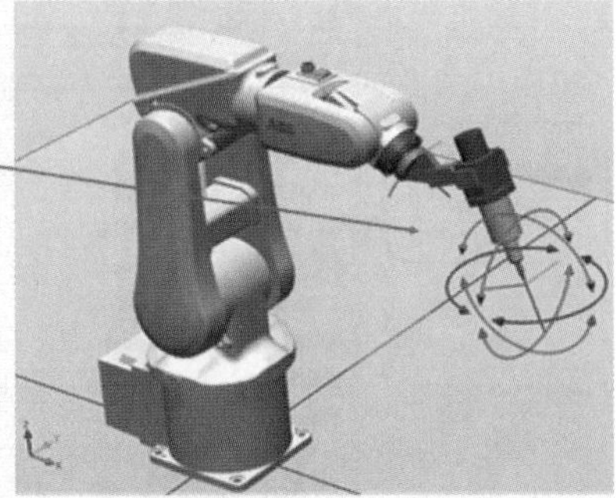

Fig.3-39 Operation Step 11

图 3-39 操作步骤 11

（11）操作示教器上的操作杆，机器人绕着工具 **TCP** 点做姿态调整的运动。

如图 **3-39** 所示。

微信扫一扫

教学视频

Lecture 3-4

Manual Operation of Industrial Robots (4)

3-4 工业机器人的手工操作（4）

1. 工业机器人转数计数器更新

ABB机器人六个关节轴都有一个机械原点的位置。在以下的情况下，需要对机械原点的位置进行转数计数器更新操作：

（1）更换伺服电机转数计数器电池后。

（2）当转数计数器发生故障，修复后。

（3）转数计数器与测量板之间断开过以后。

（4）断电后，机器人关节轴发生了位移。

（5）当系统报警提示“10036转数计数器未更新”时。

以下是进行ABB工业机器人转数计数器更新的操作。

（1）机器人六个关节轴的机械原点刻度位置示意图。

如图 3-40 所示。

1. Update of Industrial Robots’ Revolution Counter

The six joint axes of ABB robots have a mechanical origin position. In the following situations, it is necessary to update the revolution counter on the position of the mechanical origin:

(1) After the replacement of the battery of the servo motor revolution counter.

(2) After the repair of the failed revolution counter.

(3) After the disconnection of the revolution counter from the measurement board.

(4) The displacement of the joint axis of the robot after the power failure.

(5) When the system alarms that “10036 revolution counter has not been updated”.

The following is the operation to update the revolution counter of ABB industrial robot.

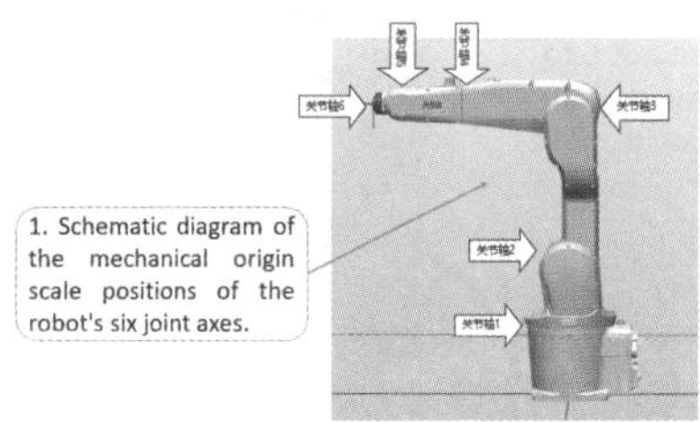

Fig.3-40 Operation Step 1

图 3-40 操作步骤 1

When manually manipulating each joint axis of an industrial robot to move to the mechanical origin scale position, it is recommended to manipulate the 4, 5, and 6 axes first, and then the 1, 2, and 3 axes. In this way, it can avoid that if 1, 2 and 3 return to the original point, the position of 4, 5 and 6 is too high, which is inconvenient for viewing and operation. The position of the mechanical origin scale of each type of industrial robot will be different.

手动操纵工业机器人各关节轴运动到机械原点刻度位置时，建议先操纵 4、5、6 轴，再操纵 1、2、3 轴，这样可以避免若 1、2、3 回到原点后，4、5、6 位置过高，不方便查看与操作。各个型号的工业机器人机械原点刻度位置会有所不同。

2. In the manual operation menu, select "Axis 4-6" as the action mode, and move the joint axis 4 to the scale position of the mechanical origin.

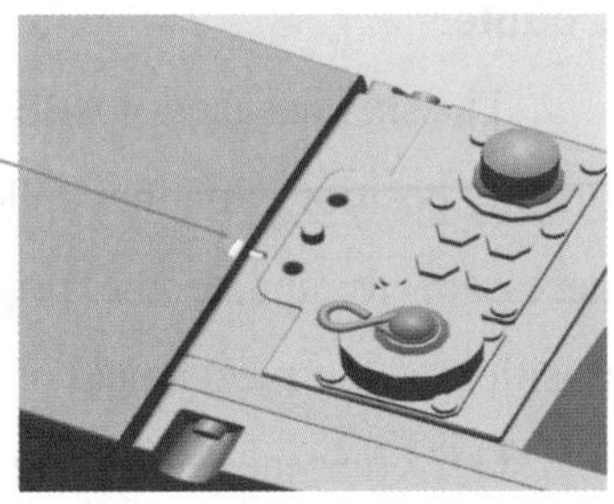

（2）在手动操纵菜单中，动作模式选择“轴 4–6”，将关节轴 4 运动到机械原点的刻度位置。

如图 3–41 所示。

Fig.3-41 Operation Step 2

图 3–41 操作步骤 2

3. In the manual operation menu, select "Axis 4-6" as the action mode, and move the joint axis 5 to the scale position of the mechanical origin.

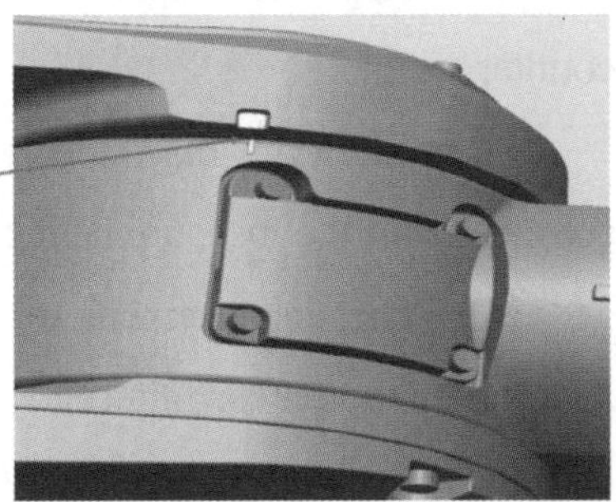

（3）在手动操纵菜单中，动作模式选择“轴 4–6”，将关节轴 5 运动到机械原点的刻度位置。

如图 3–42 所示。

Fig.3-42 Operation Step 3

图 3–42 操作步骤 3

4. In the manual operation menu, select "Axis 4-6" as the action mode, and move the joint axis 6 to the scale position of the mechanical origin.

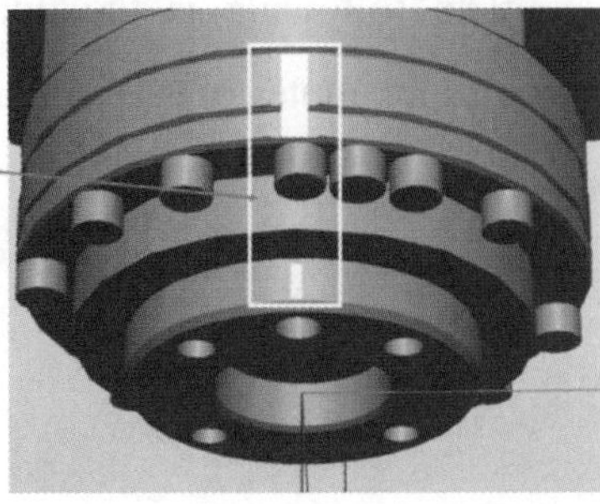

（4）在手动操纵菜单中，动作模式选择“轴 4–6”，将关节轴 6 运动到机械原点的刻度位置。

如图 3–43 所示。

Fig.3-43 Operation Step 4

图 3–43 操作步骤 4

（5）在手动操纵菜单中，动作模式选择“轴 4–6”，将关节轴 1 运动到机械原点的刻度位置。

如图 3–44 所示。

5. In the manual operation menu, select "Axis 4-6" as the action mode, and move the joint axis 1 to the scale position of the mechanical origin.

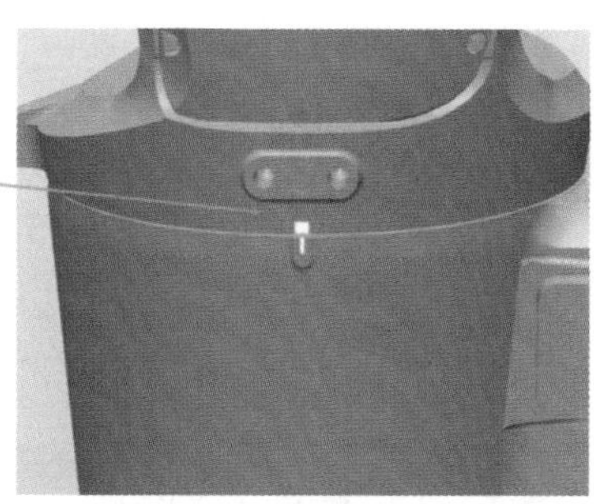

Fig.3-44 Operation Step 5

图 3–44 操作步骤 5

（6）在手动操纵菜单中，动作模式选择“轴 4–6”，将关节轴 2 运动到机械原点的刻度位置。

如图 3–45 所示。

6. In the manual operation menu, select "Axis 4-6" as the action mode, and move the joint axis 2 to the scale position of the mechanical origin.

Fig.3-45 Operation Step 6

图 3–45 操作步骤 6

（7）在手动操纵菜单中，动作模式选择“轴 4–6”，将关节轴 3 运动到机械原点的刻度位置。

如图 3–46 所示。

7. In the manual operation menu, select "Axis 4-6" as the action mode, and move the joint axis 3 to the scale position of the mechanical origin.

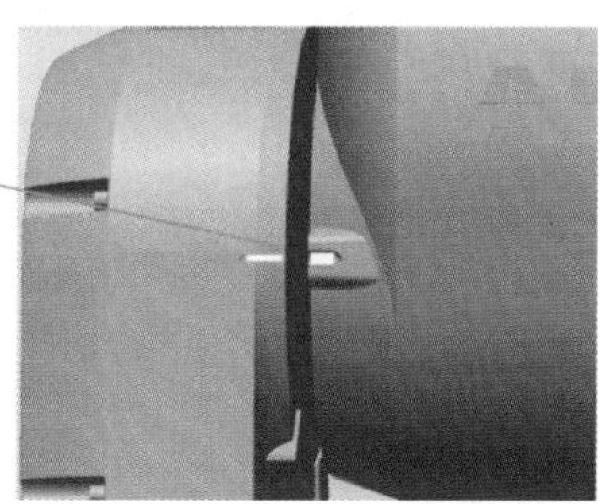

Fig.3-46 Operation Step 7

图 3–46 操作步骤 7

（8）单击左上角主菜单。

（9）选择“校准”。

如图 3–47 所示。

8. Click the main menu in the upper left corner.

9. Select "Calibration".

Fig.3-47 Operation Step 8 and 9

图 3–47 操作步骤 8 和 9

（10）单击“ROB_1”。

如图 3–48 所示。

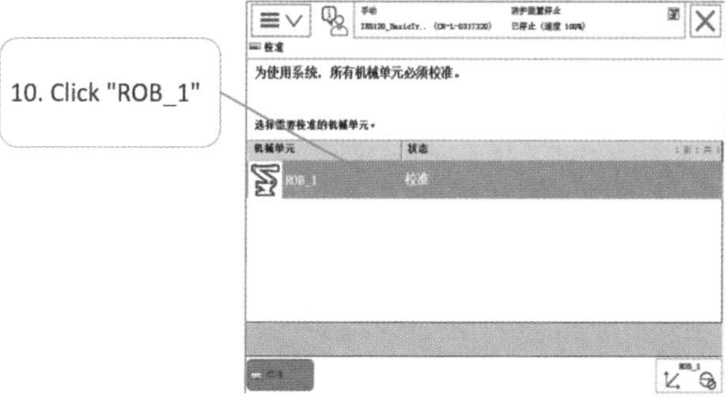

Fig.3-48 Operation Step 10

图 3–48 操作步骤 10

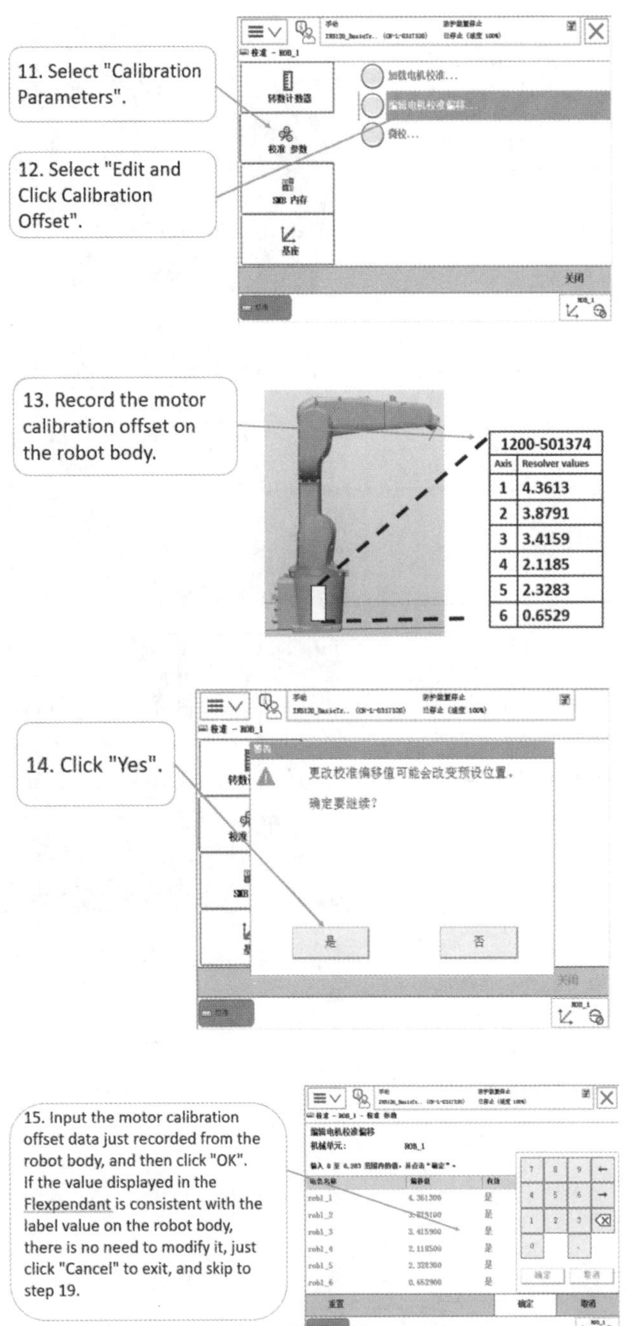

（11）选择“校准参数”。

（12）选择“编辑电机校准偏移”。

如图 3-49 所示。

Fig.3-49 Operation Step 11 and 12

图 3-49 操作步骤 11 和 12

（13）将机器人本体上电机校准偏移记录下来。

如图 3-50 所示。

Fig.3-50 Operation Step 13

图 3-50 操作步骤 13

（14）单击“是”。

如图 3-51 所示。

Fig.3-51 Operation Step 14

图 3-51 操作步骤 14

（15）输入刚才从机器人本体记录的电机校准偏移数据，然后单击“确定”。

如图 3-52 所示。

Fig.3-52 Operation Step 15

图 3-52 操作步骤 15

If the value displayed in the teaching device is consistent with the value of the label on the robot body, there is no need to modify it. Click “Cancel” to exit and skip to step 19.

如果示教器中显示的数值与机器人本体上的标签数值一致，则无须修改，直接单击“取消”退出，跳到第 19 步。

（16）单击 “是”。

如图 3-53 所示。

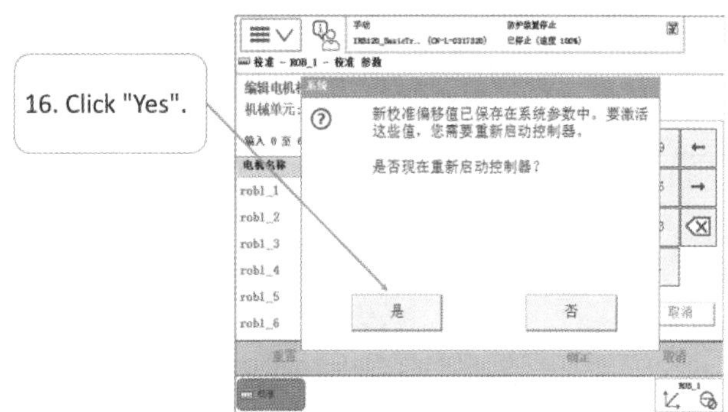

Fig.3-53 Operation Step 16

图 3-53 操作步骤 16

（17）重启后，选择“校准”。

如图 3-54 所示。

Fig.3-54 Operation Step 17

图 3-54 操作步骤 17

（18）单击 “ROB_1”。

如图 3-55 所示。

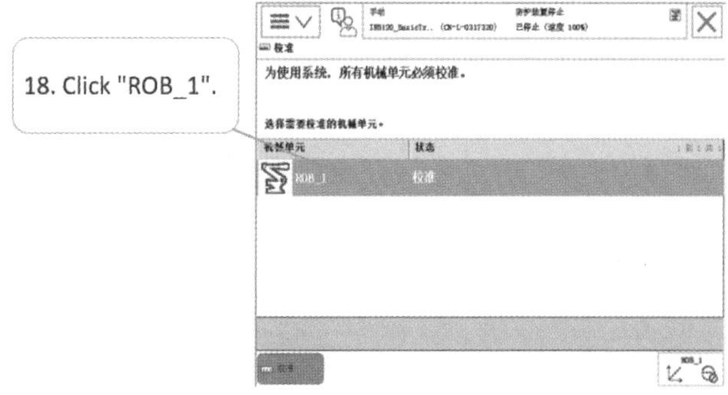

Fig.3-55 Operation Step 18

图 3-55 操作步骤 18

（19）选择 “更新转数计数器”。

如图 3-56 所示。

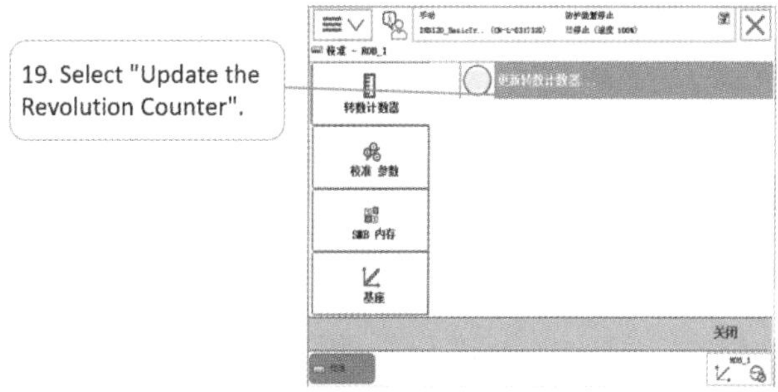

Fig.3-56 Operation Step 19

图 3-56 操作步骤 19

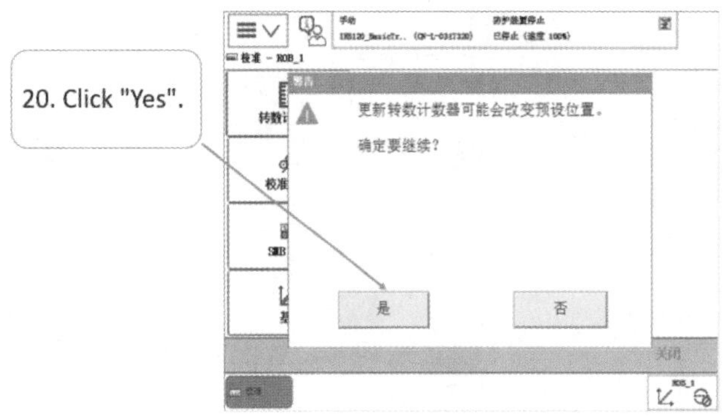

（20）单击“是”。

如图 3-57 所示。

Fig.3-57 Operation Step 20

图 3-57 操作步骤 20

21. Click "OK".

（21）单击“确定”。

如图 3-58 所示。

Fig.3-58 Operation Step 21

图 3-58 操作步骤 21

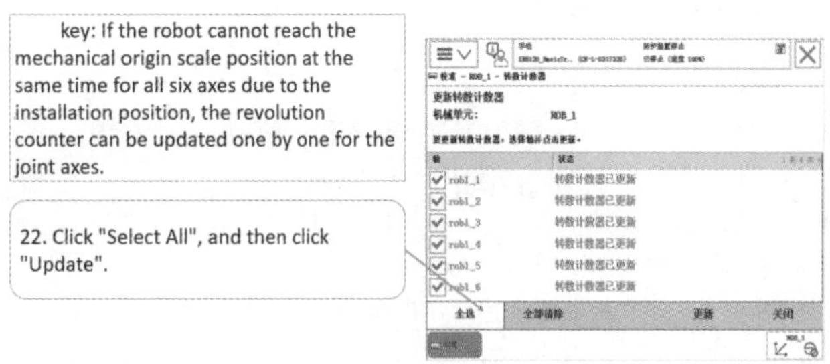

Key：如果机器人由于安装位置的关系，无法六个轴同时到达机械原点刻度位置，则可以逐一对关节轴进行转数计数器更新。

（22）单击“全选”，然后单击“更新”。

如图 3-59 所示。

Fig.3-59 Operation Step 22

图 3-59 操作步骤 22

（23）单击“更新”。

如图 3-60 所示。

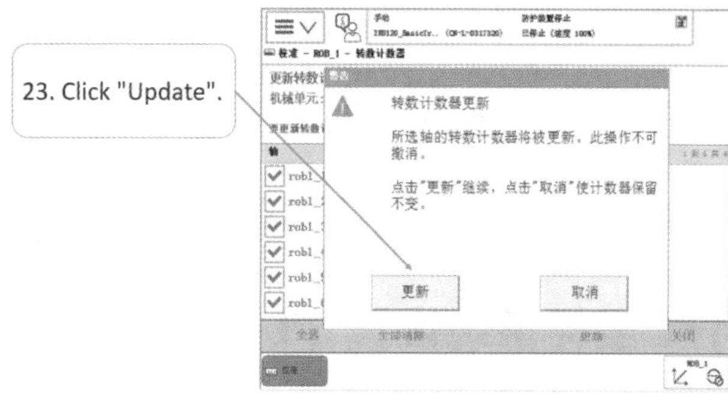

Fig.3-60 Operation Step 23

图 3-60 操作步骤 23

（24）操作完成后，转数计数器更新完成。

如图 3-61 所示。

Fig.3-61 Operation Step 24

图 3-61 操作步骤 24

微信扫一扫

教学视频

Lecture 4-1
Coordinate System Setting(1)
4-1 坐标系统设置（1）

1. Definition of Coordinate System

An industrial robot is a very complex system. In order to accurately and clearly describe the pose parameters of the robot, it is usually described by a coordinate system. The mechanism of the robot can be seen as a multi-rigid body system composed of a series of joints connected in space. Therefore, it is also a space geometry problem. The problem of space geometry is reduced to the problem of easy-to-understand algebraic form, and the method of algebra is used to calculate and prove, so as to achieve the ultimate goal of solving geometric problems.

2. Cartesian Coordinate System

In a cartesian coordinate system, the distance between a point and two mutually perpendicular coordinate axes can be used to determine the position of a point, that is, the point M in the plane corresponds to a two-digit ordered array (x, y). After the three-dimensional cartesian coordinate system is established in the space, the distance from the point to the three mutually perpendicular coordinate planes can be used to determine the position of the point, that is, the point M in the space corresponds to the three-dimensional ordered array (x, y, z). Establish a

1. 坐标系的定义

工业机器人是一个非常复杂的系统，为了准确、清楚地描述机器人位姿参数，通常采用坐标系来描述。机器人的机构可以看成一个由一系列关节连接起来的连杆在空间组成的多刚体系统，因此，也属于空间几何学问题。把空间几何学的问题归结成易于理解的代数形式的问题，用代数学的方法进行计算、证明，从而达到最终解决几何问题的目的。

2. 直角坐标系

在直角坐标系中，可用点到两条互相垂直的坐标轴的距离来确定点的位置，即平面内的点 M 与二位有序数组（x，y）一一对应。在空间建立三维直角坐标系后，可用点到三个互相垂直的坐标平面的距离来确定点的位置，即空间的点 M 与三维有序数组（x，y，z）一一对应。建立坐标系，如下图所示，取三条相互垂直的具有一定方向和度量单位的直线，叫作三维直角坐标系或空间直角坐标系 o-xyz。利用三维直角坐标系可以把空间的点M与三维有序数组（x，y，z）建立起一一对应的关系。下图 4-1 所示为典型的空间直角坐标，图 4-2 为直角坐标机器人。

coordinate system, as shown in the figure below, take three mutually perpendicular straight lines with a certain direction and measurement unit, called the three-dimensional cartesian coordinate system or the spatial cartesian coordinate system o-xyz. The three-dimensional cartesian coordinate system can establish a one-to-one correspondence between the point M in space and the three-dimensional ordered array (x, y, z). The Fig. 4-1 below shows a typical cartesian coordinate system. The Fig. 4-2 below shows a typical cartesian coordinate system robot.

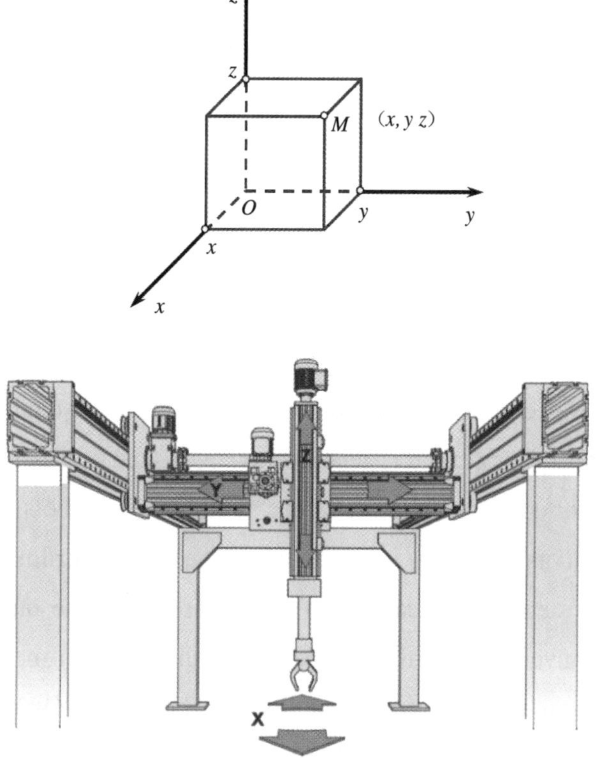

Fig.4-1 Cartesian Coordinate System in Space

图 4-1 空间直角坐标系

Fig.4-2 Cartesian Coordinate Robot

图 4-2 直角坐标机器人

以右手握住 Z 轴，当右手的四指从正向 x 轴以π/2 角度转向正向 y 轴时，大拇指的指向就是 Z 轴的正向，这样的三

Hold the Z axis with your right hand. When the four fingers of the right hand turn from the positive X axis to the y axis at an angle of π/2, the thumb is pointing to the positive direction of Z axis. Such three coordinate axes form a spatial cartesian coordinate system, and point O is called the coordinate origin. As shown in the Fig. 4-3.

条坐标轴就组成了一个空间直角坐标系，点 O 就叫做坐标原点。如图 4-3 所示。

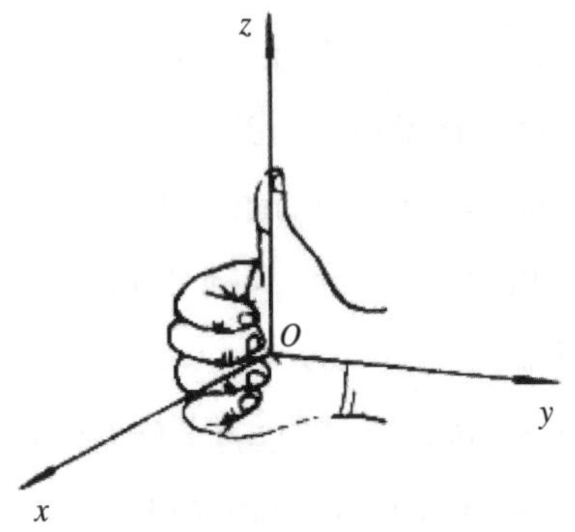

Fig.4-3 Right-handed Coordinate System
图 4-3 右手坐标系

3.Common Coordinate Systems for Industrial Robots

The commonly used coordinate systems of industrial robots are: Base Coordinate System, World Coordinate System, Work Object Coordinate System, Tool Coordinate System, and User Coordinate System. The World Coordinate System define the robot cell, and all other coordinate systems are directly or indirectly related to the World Coordinate System. It is suitable for inching control, general movement, and processing workstations and work cells with several robots or robots moved by external axes. The Base Coordinate System is located on the robot base. It is the most convenient coordinate system for the robot to move from one position to another. The Work Object Coordinate System is related to the workpiece and is usually the coordinate system most suitable for programming the robot. The Tool Coordinate System defines the position of the tool used when the robot reaches the preset target. The User Coordinate

3.工业机器人常用坐标系

工业机器人常用的坐标系有：基坐标系、大地坐标系、工件坐标系、工具坐标系、用户坐标系。大地坐标系可定义机器人单元，所有其他的坐标系均与大地坐标系直接或间接相关。它适用于微动控制、一般移动以及处理具有若干机器人或外轴移动机器人的工作站和工作单元。基坐标系位于机器人基座。它是最便于机器人从一个位置移动到另一个位置的坐标系。工件坐标系与工件相关，通常是最适于对机器人进行编程的坐标系。工具坐标系定义机器人到达预设目标时所使用工具的位置。用户坐标系在表示持有其他坐标系的设备（如工件）时非常有用。

System is very useful when representing devices (such as workpieces) that hold other coordinate systems.

（1）基坐标系

基坐标系在机器人基座中有相应的零点，这使固定安装的机器人的移动具有可预测性。因此它对于将机器人从一个位置移动到另一个位置很有帮助。对机器人编程来说，其他如工件坐标系等坐标系通常是最佳选择。在正常配置的机器人系统中，当您站在机器人的前方并在基坐标系中微动控制，将控制杆拉向自己一方时，机器人将沿 X 轴移动；向两侧移动控制杆时，机器人将沿 Y 轴移动。扭动控制杆，机器人将沿 Z 轴移动。如下图 4-4 所示为机器人基坐标系。

(1) Base Coordinate System

The Base Coordinate System has a corresponding zero point in the robot base, which makes the movement of the fixedly installed robot predictable. So it is very helpful for moving the robot from one position to another. For robot programming, other coordinate systems such as the Work Object Coordinate System are usually the best choice. In a normally configured robot system, the robot will move along the X axis when you stand in front of the robot and do the inching control in the Base Coordinate System, pulling the joystick to your side; when you move the joystick to both sides, the robot will move along the Y axis. Twist the joystick and the robot will move along the Z axis. The following Fig. 4-4 shows the robot Base Coordinate System.

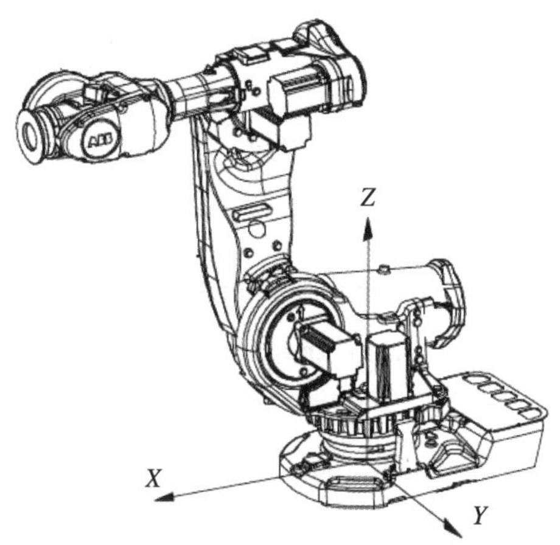

Fig.4-4 Base Coordinate System of Industrial Robot

图 4-4 工业机器人基坐标系

（2）大地坐标系

大地坐标系在工作单元或工作站中的固定位置有其相应的零点。这有助于处理若干个

(2) World Coordinate System

The World Coordinate System has its correspondin g zero point at a fixed position in the work cell or workstation. This helps to handle several robots or robots that move by external axes. By default, the World Coordinate System is consistent with the Base Coordinate System. Suppose there are two robots, one is installed on the ground and the other is upside down. The Base Coordinate System of the inverted robot will also be upside down. If inching control is performed in the Base Coordinate System of the inverted robot, it is difficult to predict the movement. At this time, you can choose the shared World Coordinate System to replace. As shown in the Fig. 4-5.

机器人或由外轴移动的机器人。在默认情况下，大地坐标系与基坐标系是一致的。假如，有两个机器人，一个安装于地面，一个倒置。倒置机器人的基坐标系也将上下颠倒。如果您在倒置机器人的基坐标系中进行微动控制，则很难预测移动情况。此时可选择共享大地坐标系取而代之。如图 4-5 所示。

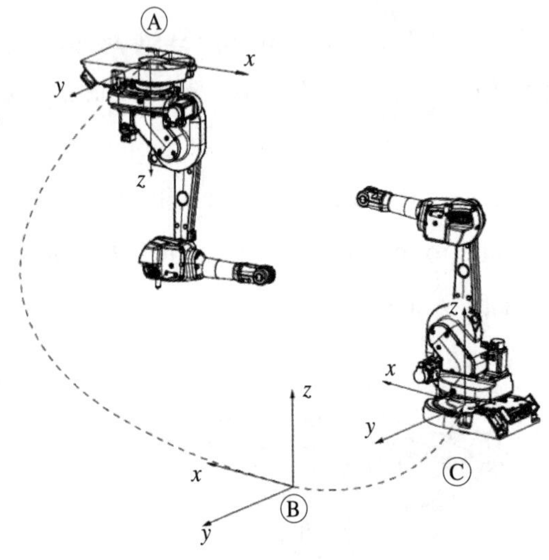

Fig.4-5 World Coordinate System of Industrial Robot
A Base Coordinate System of Robot 1
B World Coordinate System
C Base Coordinate System of Robot 2

图 4-5 工业机器人大地坐标系
A 机器人 1 基坐标系 B 大地坐标系
C 机器人 2 基坐标系

(3) Tool Coordinate System

The Tool Coordinate System is composed of the tool center point (TCP) and the coordinate axis orientation. When moving, the TCP will move strictly according to the specified path and speed of the program. All robots have a predefined Tool Coordinate System at the wrist, and the default tool center point of tool zero is located at the center of the 6-axis. In this way, one or more new Tool Coordinate Systems

（3）工具坐标系

工具坐标系是由工具中心点（TCP）与坐标轴方位构成的，运动时 TCP 会严格按程序指定路径和速度运动。所有机器人在手腕处都有一个预定义工具坐标系，默认工具 tool0 中心点位于 6 轴中心，这样就能将一个或多个新工具坐标系定义为 tool0 的偏移值。机器人联动运行时，TCP 是必需的，程序中支持多个工具，可根据当前工作状态进行变换，比如焊接程序可以定义多个工具对应不同的延伸长度，工具被更换之后，重新定义工具即可直接运行程序。下图 4-6 为工业机器人工具坐标系。

can be defined as the offset value of tool zero. When the robot is running in linkage, TCP is required. The program supports multiple tools, which can be changed according to the current working status. For example, a welding program can define multiple tools corresponding to different extension lengths. After the tool is replaced, redefine the tool to run the program directly. The Tool Coordinate System of industrial robots is shown in Fig.4-6 below.

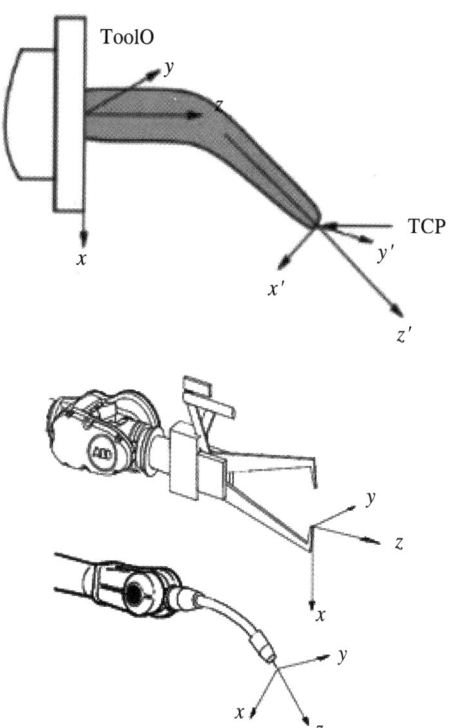

Fig.4-6　Tool Coordinate System of Industrial Robot

图 4-6　工业机器人的工具坐标系

（4）工件坐标系

工件坐标对应工件，它定义工件相对于大地坐标的位置。工业机器人可以有若干工件坐标系，或者表示不同工件，或者表示同一工件在不同位置的若干副本。对机器人进行编

(4) Work Object Coordinate System

The Work Object Coordinate corresponds to the workpiece, which defines the position of the workpiece relative to the World Coordinate. Industrial robots can have several Work Object Coordinate Systems, represent different workpieces, or represent several copies of the same workpiece at different positions.

Programming the robot is to create targets and paths in the Work Object Coordinates, which brings the following advantages:

程就是在工件坐标中创建目标和路径，这带来以下优点：

①When repositioning the workpiece in the workstation, you only need to change the position of the Work Object Coordinate, and all paths will be updated immediately.

①当重新定位工作站中的工件时，只需更改工件坐标的位置，所有路径将即刻随之更新。

② It is allowed to operate the workpiece that is moved by an external axis or conveying rail, because the entire workpiece can move along with its path. Fig. 4-7 shows the coordinate system of industrial robot workpiece.

②允许操作以外部轴或传送导轨移动的工件，因为整个工件可连同其路径一起移动。图 4-7 所示为工业机器人工件坐标系。

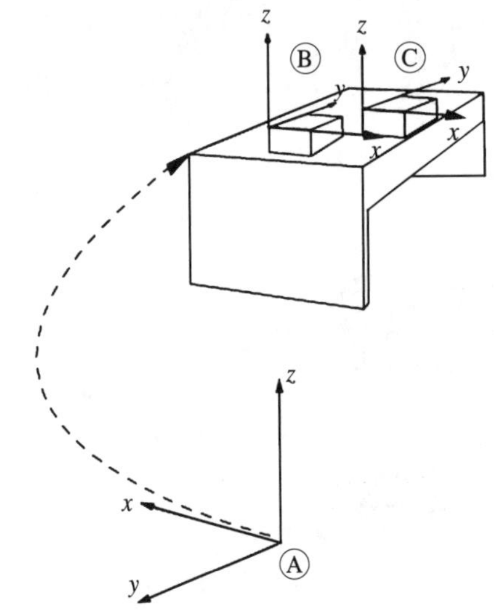

Fig.4-7 Work Object Coordinate System of Industrial Robot

图 4-7 工业机器人工件坐标系

(5) Coordinate System of Users

（5）用户坐标系

The User Coordinate System can be used to represent fixtures, worktables and other equipment. This provides an extra level in the relevant coordinate system chain, which helps to handle processing equipment holding workpieces or other coordinate systems. The figure shows several commonly-used coordinate systems. As shown in the Fig. 4-8.

用户坐标系可用于表示固定装置、工作台等设备。这就在相关坐标系链中提供了一个额外级别，有助于处理持有工件或其他坐标系的处理设备。图 4-8 中表示了几种常用的坐标系。

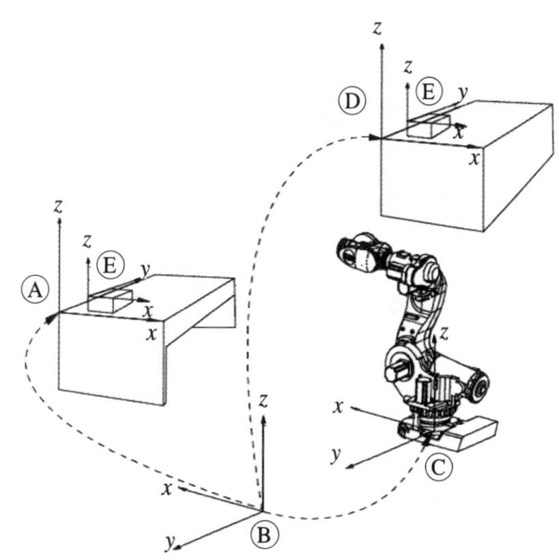

Fig. 4-8 Several Commonly-used Coordinate Systems
A Coordinate System of Users
B World Coordinate System
C Base Coordinate System
D Coordinate System of Mobile Users
E Work Object Coordinate System, move together with the coordinate system of users
图 4-8 几种常用的坐标系
A 用户坐标系 B 大地坐标系
C 基坐标系 D 移动用户坐标系
E 工件坐标系，与用户坐标系一同移动

微信扫一扫

教学视频

Lecture 4-2
Coordinate System Setting(2)

4-2 坐标系统设置（2）

1. Tool Data “tooldata”

Tool data is used to describe the parameter data such as the tool coordinate TCP (the origin of the Tool Coordinate System, also is the tool center point) installed on the sixth axis of the robot, mass, and center of gravity. Tool data will affect the robot's control algorithm (such as calculating acceleration), speed and acceleration monitoring, torque monitoring, collision monitoring, energy monitoring, etc. Therefore, the tool data of robot needs to be set correctly. All robots have a predefined tool coordinate system at the wrist, which is called tool zero. In this way, one or more new Tool Coordinate Systems can be defined as the offset value of tool zero.

The tool center point of the default tool (tool zero) is located at the center of the robot mounting flange. When the program is executed, the robot moves the TCP to the programmed position, as shown in Figure 4-9 below.

1. 工具数据 “tooldata”

工具数据 tooldata 用于描述安装在机器人第六轴上的工具坐标 TCP（也就是工具坐标系的原点即工具中心点）、质量、重心等参数数据。工具数据 tooldata 会影响机器人的控制算法（例如计算加速度）、速度和加速度监控、力矩监控、碰撞监控、能量监控等，因此机器人的工具数据需要正确设置。所有机器人在手腕处都有一个预定义的工具坐标系，该坐标系被称为 tool0。这样就能将一个或者多个新工具坐标系定义为 tool0 的偏移值。

默认工具（tool0）的工具中心点位于机器人安装法兰的中心，执行程序时，机器人将 TCP 移至编程位置，如下图 4-9 所示。

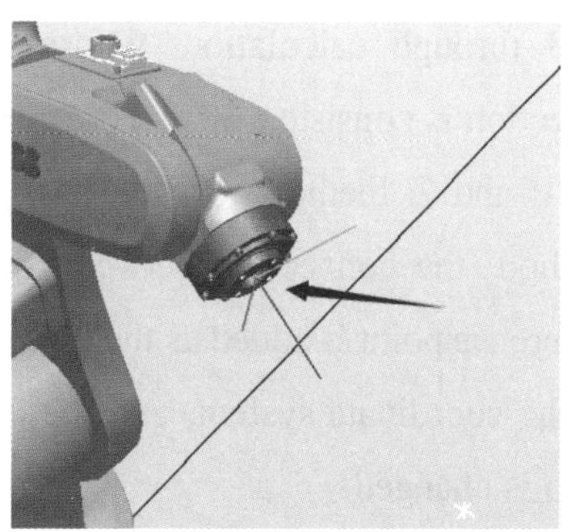

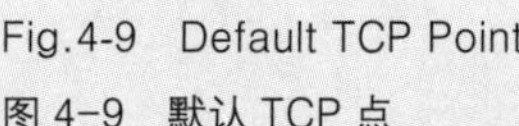
Fig.4-9 Default TCP Point
图 4-9 默认 TCP 点

一般，不同的机器人应用就可能配置不同的工具，比如说弧焊的机器人就使用弧焊枪作为工具，而用于搬运板材的机器人就会使用吸盘式的夹具作为工具。如下图 4-10 所示为不同工具的 TCP 点。

Generally, different robot applications may be equipped with different tools. For example, an arc welding robot uses an arc welding torch as a tool, while a robot used to handle plates uses a suction cup fixture as a tool. The TCP points of different tools are shown in Fig. 4-10 below.

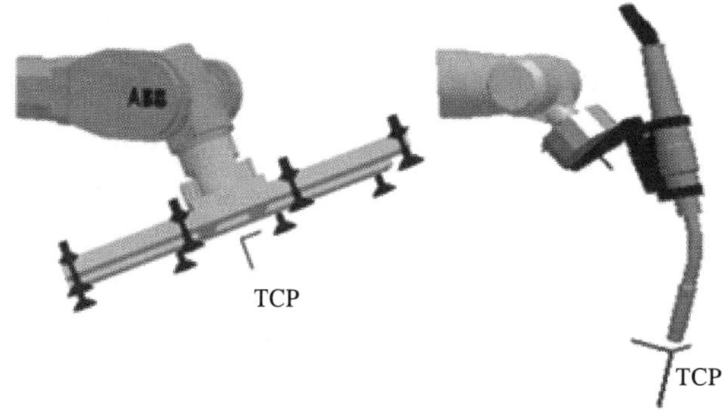

Fig.4-10 TCP of Different Tools
图 4-10 不同工具的 TCP

TCP 的设定方法包括 N（3≤N≤9）点法、TCP 和 Z 法、TCP 和 Z，X 法。

The TCP setting methods include N (3≤N≤9) points method, TCP and Z method, TCP and Z, X method.

（1）N（3≤N≤9）点法：机器人的 TCP 通过 N 种不同的姿态同参考点接触，得出多组解，通过计算得出当前 TCP 与机器人安装法兰中心点

(1)N (3≤N≤9) points method: The TCP of the robot contacts the reference point through N different postures to obtain multiple sets of solutions. The corresponding position of the current TCP and the center point of the robot mounting flange (Tool zero)

is obtained through calculation, whose coordinate system direction is consistent with Tool zero.

(2)TCP and Z method: On the basis of the N points method, the connection between the Z point and the reference point is added as the direction of the Z axis of the coordinate system, and the Z direction of tool zero is changed.

(3)TCP and Z, X method: On the basis of the N point method, the connection between the X point and the reference point is added as the direction of the X axis of the coordinate system, and the connection between the Z point and the reference point is the direction of the Z axis of the coordinate system. The X and Z directions of tool zero are changed.

The method of setting tooldata usually adopts TCP and Z, X method (N=4). The setting principle is:

(1)First, find a very accurate fixed point as a reference point within the working range of the robot;

(2)Then determine a reference point on the tool (preferably the center point of the tool);

(3)Manipulate the robot manually to move the reference point on the tool, and use more than four different robot postures to just hit the fixed point as much as possible. The posture difference of the first three points is as large as possible, which is conducive to the improvement of TCP accuracy. In order to obtain a more accurate TCP, the six-point method is used in the following example, which is TCP and Z, X method (N=4). The tool's reference point perpendicular to the fixed point is used as the fourth point, and the fifth point is the tool reference point

（Tool0）相应位置，其坐标系方向与 Tool0 一致。

（2）TCP 和 Z 法：在 N 点法基础上，增加 Z 点与参考点的连线为坐标系 Z 轴的方向，改变了 tool0 的 Z 方向。

（3）TCP 和 Z，X 法：在 N 点法基础上，增加 X 点与参考点的连线为坐标系 X 轴的方向，Z 点与参考点的连线为坐标系 Z 轴的方向，改变了 tool0 的 X 和 Z 方向。

设定工具数据 tooldata 的方法通常采用 TCP 和 Z，X 法（N=4）。其设定原理如下：

（1）首先在机器人工作范围内找一个非常精确的固定点作为参考点；

（2）然后在工具上确定一个参考点（最好是工具的中心点）；

（3）用手动操纵机器人的方法，去移动工具上的参考点，以四种以上不同的机器人姿态尽可能与固定点刚好碰上。前三个点的姿态相差尽量大些，这样有利于 TCP 精度的提高。为了获得更准确的 TCP，在以下的例子中使用六点法也就是 TCP 和 Z，X 法（N=4）进行操作，第四点是用工具的参考点垂直于固定点，第五点是工具参考点从固定点向将要设定为 TCP 的 X 方向移动，第六点是工具参考点从固定点向将要设定为 TCP 的 Z 方向移动；

（4）机器人通过这四个位置点的位置数据计算求得 TCP 的数据，然后 TCP 的数据就保存在 tooldata（这个程序数据）中被程序进行调用。

2. 工件数据 wobjdata

工件坐标系对应工件，它定义工件相对于大地坐标系（或其他坐标系）的位置。机器人可以拥有若干工件坐标系，或者表示不同工件，或者表示同一工件在不同位置的若干副本。对机器人进行编程时在工件坐标系中创建目标和路径，将产生很多优点。

（1）重新定位工作站中的工件时，只需更改工件坐标系的位置，所有路径将即刻随之更新。

（2）允许操作以外轴或传送导轨移动的工件，因为整个工件可连同其路径一起移动。

下图 4-11 中，A 是机器人的大地坐标，为了方便编程为第一个工件建立了一个工件坐标 B，并在这个工件坐标 B 进行轨迹编程。如果台子上还有一个一样的工件需要走一样

moving from the fixed point to the X direction to be set as TCP. The sixth point is the tool reference point to move from the fixed point to the Z direction to be set as TCP;

(4)The robot calculates the TCP data through the position data of these four position points, and then the TCP data is saved in the tooldata (the program data) and called by the program.

2. Workpiece Data wobjdata

The Work Object Coordinate System corresponds to the workpiece, which defines the position of the workpiece relative to the World Coordinate System (or other coordinate system). The robot may have several Work Object Coordinate Systems, representing different workpieces, or several copies of the same workpiece at different positions. Creating targets and paths in the Work Object Coordinate System when programming the robot will yield many advantages.

(1) When repositioning the workpiece in the workstation, you only need to change the position of the Work Object Coordinate System, and all paths will be updated immediately.

(2) It is allowed to operate the workpiece moved by the external axis or the conveying rail, because the entire workpiece can move along with its path.

In Fig. 4-11 below, A is the World Coordinate System of the robot. For the convenience of programming, a workpiece coordinate B is established for the first workpiece, and the trajectory programming is performed on this work object coordinate B. If there is a same workpiece on the table that needs to follow the same trajectory, then you only need to establish a work object coordinate C, copy the trajectory in the work object coordinate B, and then update the work object coordinate from B to C. There is no need to program the same workpiece for repeated trajectories.

的轨迹，那你只需要建立一个工件坐标 C，将工件坐标 B 中的轨迹复制一份，然后将工件坐标从 B 更新为 C，则无需对一样的工件重复的轨迹编程了。

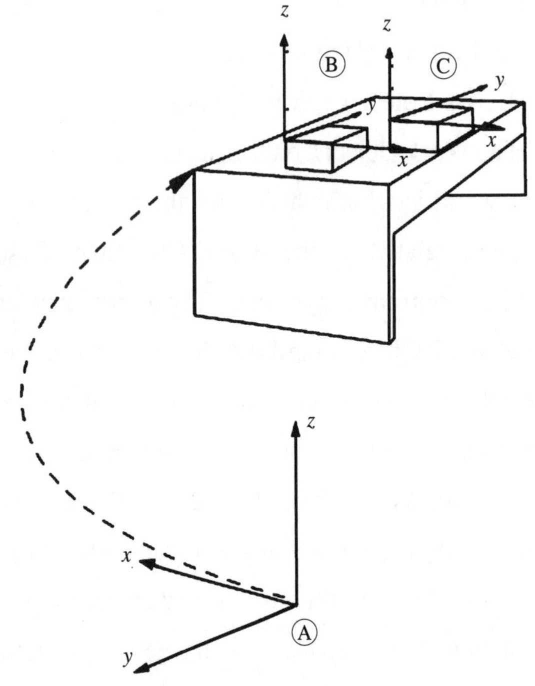

Fig.4-11 Application of Work Object Coordinate System
图 4-11 工件坐标系应用

In Fig. 4-12 below, the trajectory programming of object A is carried out in work object coordinate B. If the position of the work object coordinate changes to the work object coordinate D, you only need to redefine the work object coordinate D in the robot system, and the robot's trajectory will be automatically updated to C, and there is no need to reprogram the trajectory. Because the relationship between A relative to B and C relative to D is the same, there is no change due to the overall offset.

下图 4-12 中，在工件坐标 B 中对 A 对象进行了轨迹编程。如果工件坐标的位置变化成工件坐标 D 后，只需在机器人系统重新定义工件坐标 D，则机器人的轨迹就自动更新到 C 了，不需要再次轨迹编程了。因 A 相对于 B，C 相对于 D 的关系是一样，并没有因为整体偏移而发生变化。

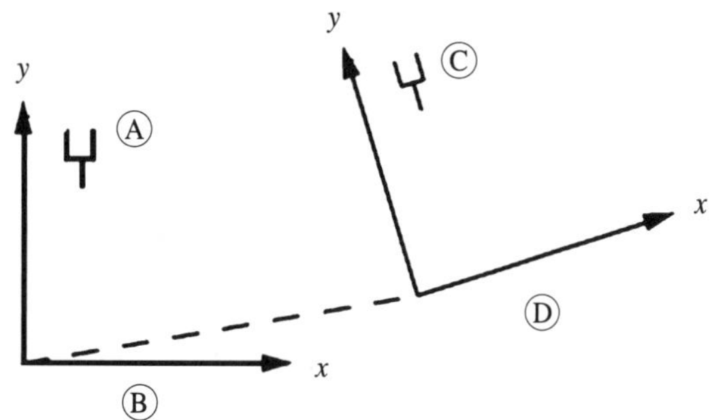

Fig.4-12 Application of Work Object Coordinate System
图 4-12 工件坐标系应用

工件坐标系设定时，通常采用三点法。只需在对象表面位置或工件边缘角位置上，定义三个点位置，来创建一个工件坐标系。其设定原理如下：

（1）手动操纵机器人，在工件表面或边缘角的位置找到一点 X1，作为坐标系的原点；

（2）手动操纵机器人，沿着工件表面或边缘找到一点 X2，X1、X2 确定工件坐标系的 X 轴的正方向，（X1 和 X2 距离越远，定义的坐标系轴向越精准）；

（3）手动操纵机器人，在 XY 平面上并且 Y 值为正的方向找到一点 Y1，确定坐标系的 Y 轴的正方向。

The three-point method is usually used when setting the Work Object Coordinate System. Just define three point positions on the surface position of the object or the corner position of the workpiece edge to create a Work Object Coordinate System. The setting principle is as follows:

(1) Manually manipulate the robot to find a point X1 at the position of the workpiece surface or edge corner as the origin of the coordinate system;

(2) Manually manipulate the robot to find a point X2 along the surface or edge of the workpiece. X1 and X2 determine the positive direction of the X axis of the Work Object Coordinate System. (The greater the distance between X1 and X2 is, the more precise the axis of the defined coordinate system is);

(3) Manually manipulate the robot, find a point Y1 on the XY plane with a positive Y value, and determine the positive direction of the Y axis of the coordinate system.

微信扫一扫

教学视频

Lecture 4-3

Coordinate System Setting(3)

4-3 坐标系统设置（3）

1. Setting of Tool Coordinate System

TCP setting methods include N (N≥3) point method, TCP and Z method, TCP and Z, X method.

(1) N (N≥3) point method. The TCP of the robot contacts the reference point through N different postures, and multiple sets of solutions are obtained. The relative position of the current TCP and the center point of the robot mounting flange (Tool0) is obtained by calculation, and the coordinate system direction is consistent with Tool0.

(2) TCP and Z method. On the basis of the N point method, the line connecting the Z point and the reference point is the direction of the Z axis of the coordinate system.

(3) TCP and Z,X methods. On the basis of the N point method, the connection line between the X point and the reference point is the direction of the X axis of the coordinate system, and the connection line between the Z point and the reference point is the direction of the Z axis of the coordinate system.

The operation method of establishing a new tool data tool1 with TCP and Z, X method (N=4) is as follows:

1. 工具坐标系的设定

TCP 的设定方法包括 N（N≥3）点法，TCP 和 Z 法，TCP 和 Z、X 法。

（1）N（N≥3）点法。机器人的 TCP 通过 N 种不同的姿态同参考点接触，得出多组解，通过计算得到当前 TCP 与机器人安装法兰中心点（Tool0）的相对位置，其坐标系方向与 Tool0 一致。

（2）TCP 和 Z 法。在 N 点法基础上，Z 点与参考点连线为坐标系 Z 轴的方向。

（3）TCP 和 Z、X 法。在 N 点法基础上，X 点与参考点连线为坐标系 X 轴的方向，Z 点与参考点连线为坐标系 Z 轴的方向。

以 TCP 和 Z、X 法（N=4）建立一个新的工具数据 tool1 的操作方法如下：

（1）单击“ABB”按钮，弹出如右图所示窗口。如图4–13所示。

(1) Click the “ABB” button, and the window shown above will pop up. As it is shown in Fig. 4-13.

Fig.4-13 Operation Step 1

图4–13 操作步骤1

（2）如图4–14所示，选择“手动操作。”

(2) As shown in Fig. 4-14, select “Manual Operation.”

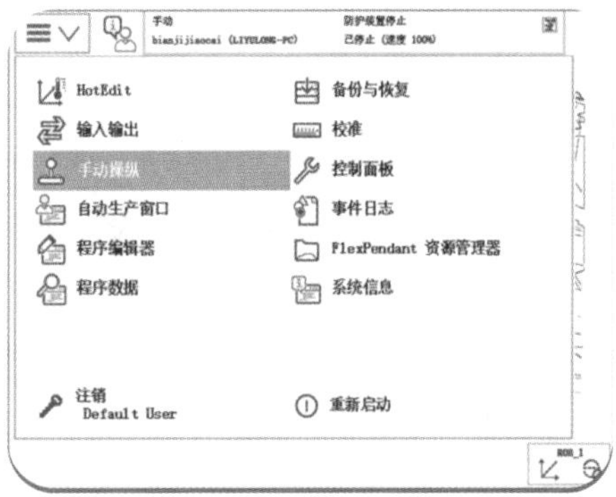

Fig.4-14 Operation Step 2

图4–14 操作步骤2

（3）如图4–15所示，选择“工具坐标”。

(3) As shown in Fig. 4-15, select “Tool Coordinates”.

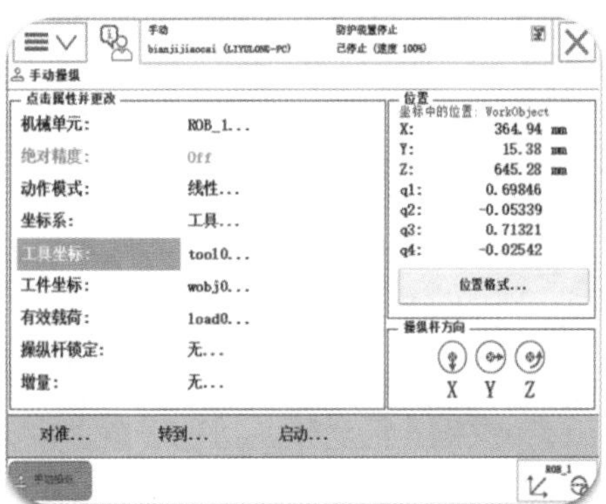

Fig.4-15 Operation Step 3

图4–15 操作步骤3

(4) As shown in Fig.4-16, click "New…".

（4）如图 4−16 所示，单击“新建”。

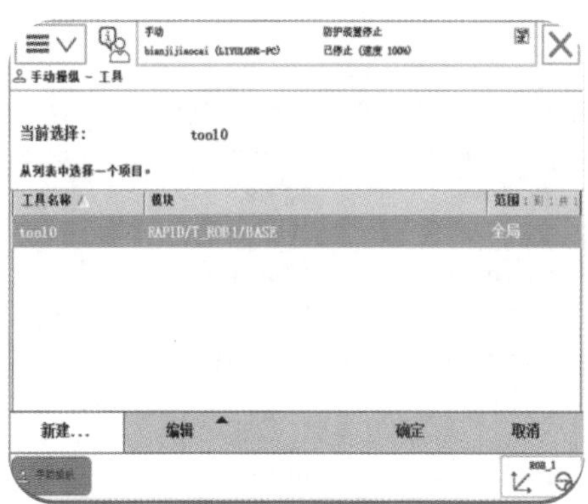

Fig.4-16 Operation Step 4

图 4−16 操作步骤 4

(5) As shown in Fig.4-17, select tool1 and click the "Define…" option in the "Edit" menu.

（5）如图 4−17 所示，选中 tool1，单击“编辑”菜单中的“定义”选项。

Fig.4-17 Operation Step 5

图 4−17 操作步骤 5

(6) As shown in Fig.4-18, select "TCP and Z, X" and the number of points N=4 to set TCP.

（6）如图 4−18 所示，选择“TCP 和 Z，X”，点数 N=4 来设定 TCP。

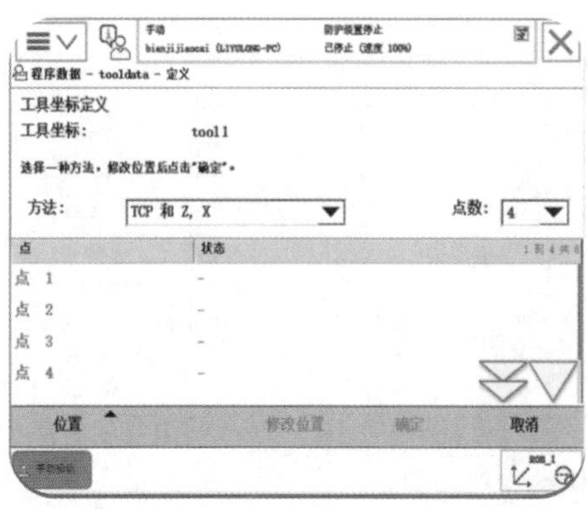

Fig.4-18 Operation Step 6

图 4−18 操作步骤 6

（7）如图 4−19 所示，通过示教器选择合适的手动操纵模式。

(7) As shown in Fig. 4-19, select the appropriate manual operation mode through the Flexpendant.

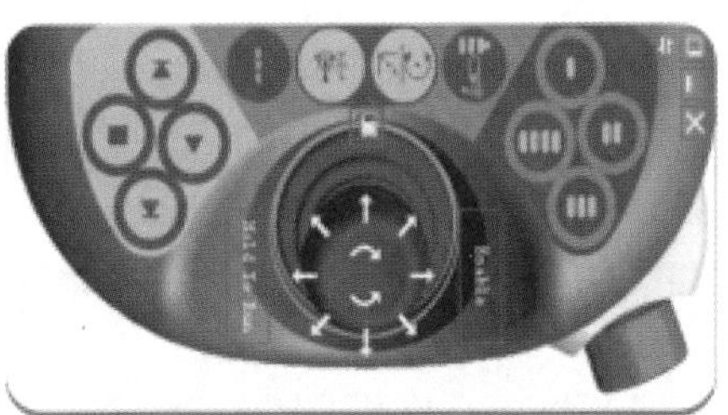

Fig.4-19 Operation Step 7
图 4−19 操作步骤 7

（8）如图 4−20 所示，按下使能键，操作手柄靠近固定点，如右图所示机器人姿势作为第一个点，单击“修改位置”完成第一个点的修改。

(8) As shown in Fig. 4-20, Press the enabler key, the operating handle is close to the fixed point, as shown in the figure above, the robot posture is taken as the first point, click "Modify Position" to complete the modification of the first point.

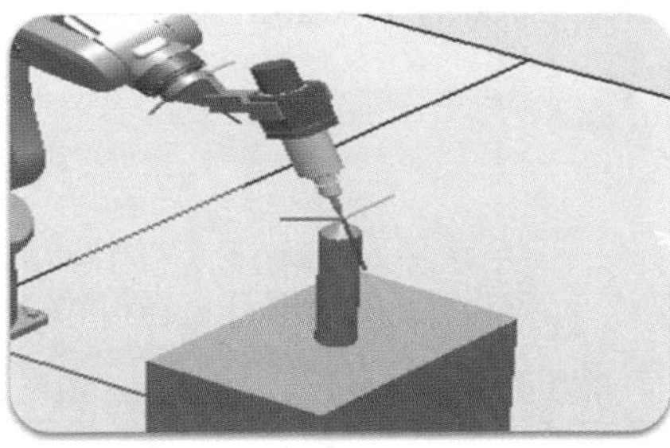

Fig.4-20 Operation Step 8
图 4−20 操作步骤 8

（9）按照上面的操作依次完成对点 2、3、4 的修改；点 2、点 3、点 4 对应的机器人姿态如图 4−21、图 4−22、图 4−23 所示。

(9) Follow the above operations to complete the modifications to points 2, 3, and 4 in turn; Robot poses corresponding to points 2, 3 and 4 are shown in Fig. 4-21、4-22、4-23.

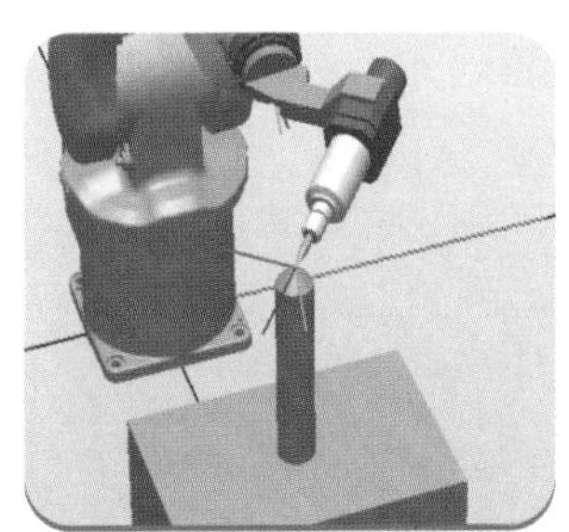

Fig.4-21 Operation Step 9
图 4−21 操作步骤 9

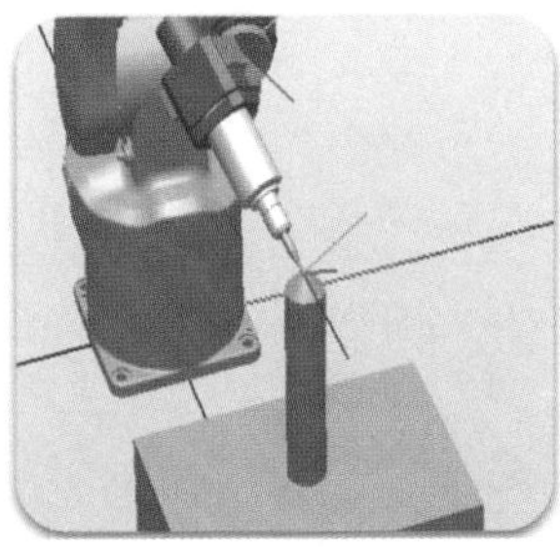

Fig.4-22 Operation Step 10
图 4−22 操作步骤 10

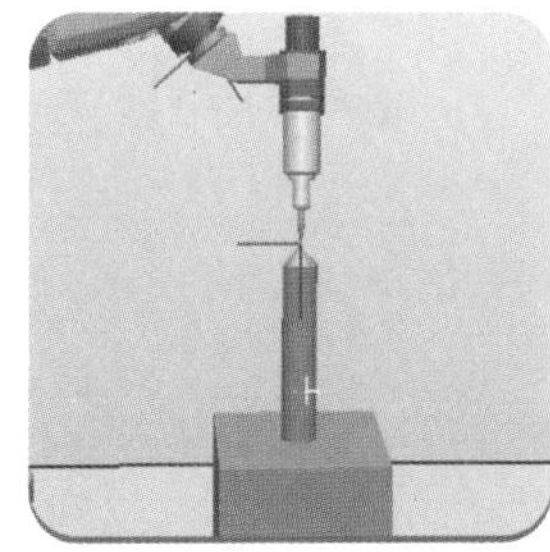

Fig.4-23 Operation Step 11
图 4−23 操作步骤 11

The four points are modified as shown in Fig. 4-24.

四个点修改完成后如图4-24所示。

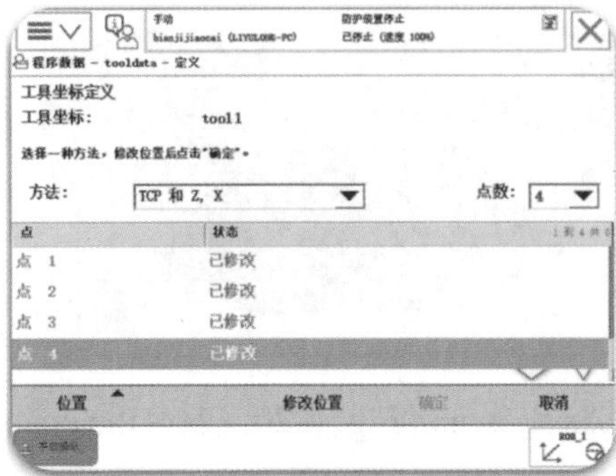

Fig.4-24 Operation Step 12
图 4-24 操作步骤 12

(10) As shown in Fig.4-25, manipulate the robot to move the tool reference point from the fixed point to the +X direction of the tool TCP in the posture of point 4, click "Modify Position".

（10）如图4-25所示，操控机器人使工具参考点以点4的姿态从固定点移动到工具TCP的+X方向。如左图所示，单击“修改位置”。

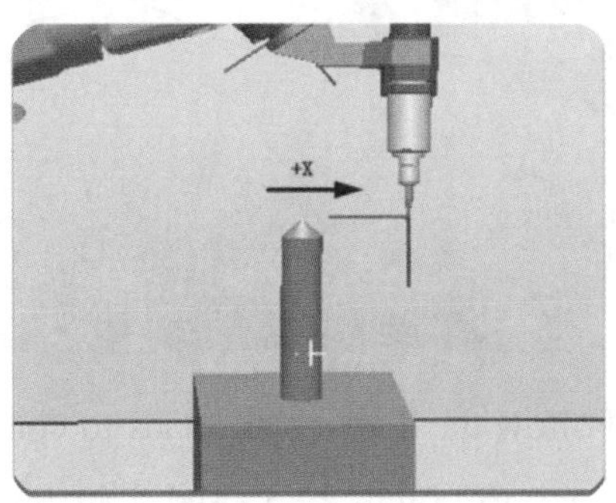

Fig.4-25 Operation Step 13
图 4-25 操作步骤 13

(11) As shown in Fig.4-26, manipulate the robot to move the tool reference point from the fixed point to the +Z direction of the tool TCP in the posture of point 4, click "Modify Position".

（11）如图4-26所示，操控机器人使工具参考点以点4的姿态从固定点移动到工具TCP的+Z方向，单击“修改位置”。

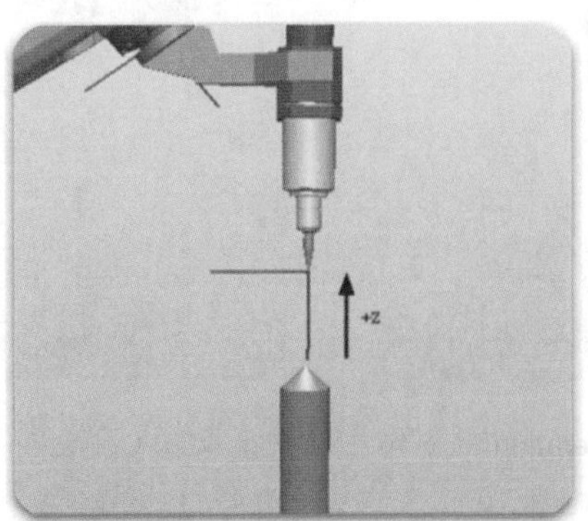

Fig.4-26 Operation Step 14
图 4-26 操作步骤 14

（12）如图 4-27 所示，单击“确定”完成位置修改。

(12) As shown in Fig. 4-27, click “OK” to complete the position modification.

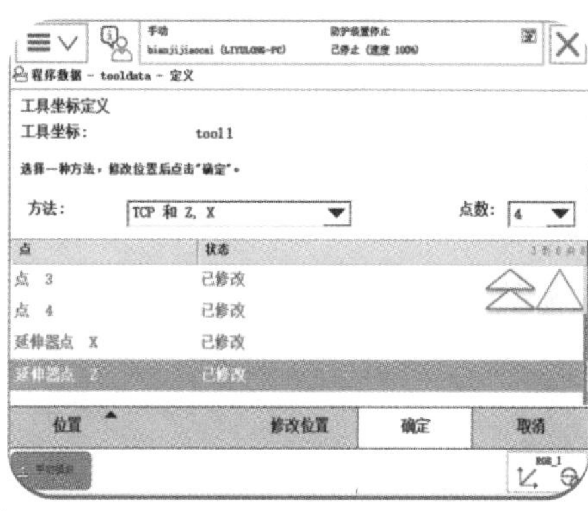

Fig.4-27 Operation Step 15
图 4-27 操作步骤 15

（13）如图 4-28 所示，查看误差，越小越好，但也要以实际验证效果为准。

(13) As shown in Fig.4-28, check the error, the smaller the better, but the actual verification effect shall prevail.

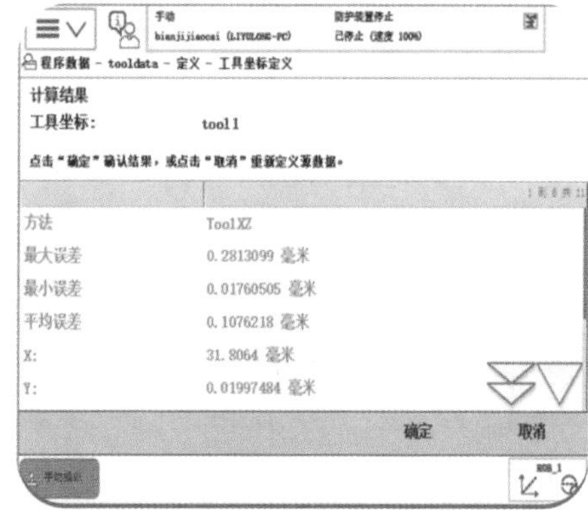

Fig.4-28 Operation Step 16
图 4-28 操作步骤 16

（14）选中“tool1”，如图 4-29 所示，然后打开编辑菜单选择“更改值”。

(14) Select “tool1”, as shown in the Fig.4-29, then open the edit menu and select “Change Value”.

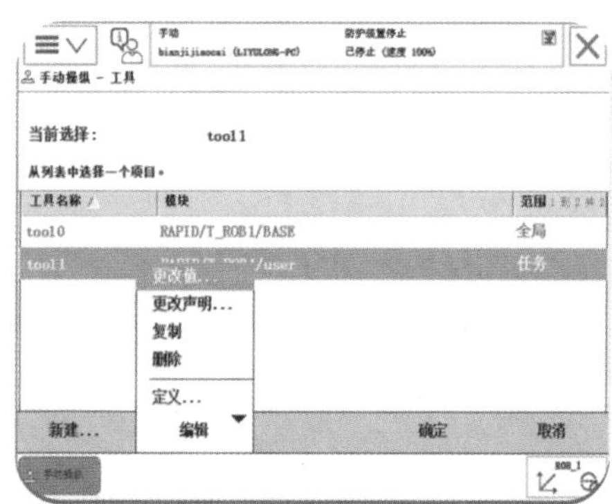

Fig.4-29 Operation Step 17
图 4-29 操作步骤 17

(15) As shown in Fig.4-30 is the value change menu of tool1.

（15）如图 4-30 所示为 tool1 的更改值菜单。

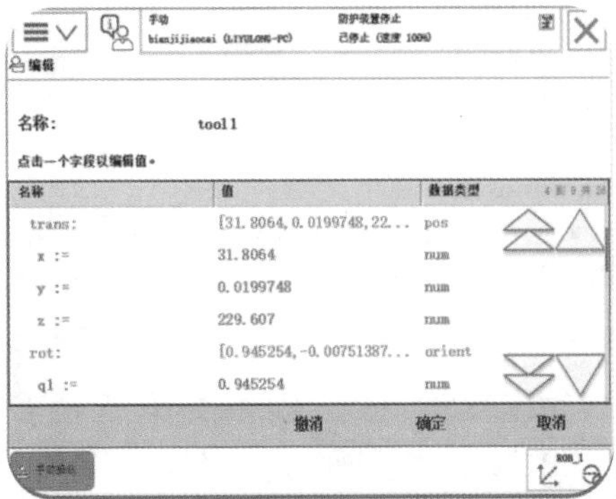

Fig.4-30 Operation Step 18

图 4-30 操作步骤 18

(16) As shown in Fig.4-31, Click the arrow to scroll down and change the value of mass to the actual weight of the tool (unit: kg).

（16）如图 4-31 所示，单击箭头向下翻页，将 mass 的值改为工具的实际重量（单位 kg）。

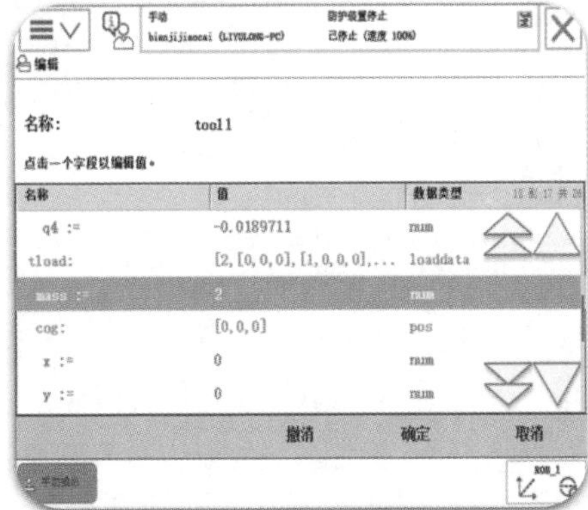

Fig.4-31 Operation Step 19

图 4-31 操作步骤 19

(17) As shown in Fig.4-32, the coordinates of the center of gravity of the editing tool are best based on actual conditions.

（17）如图 4-32 所示，编辑工具重心坐标，以实际为准最佳。

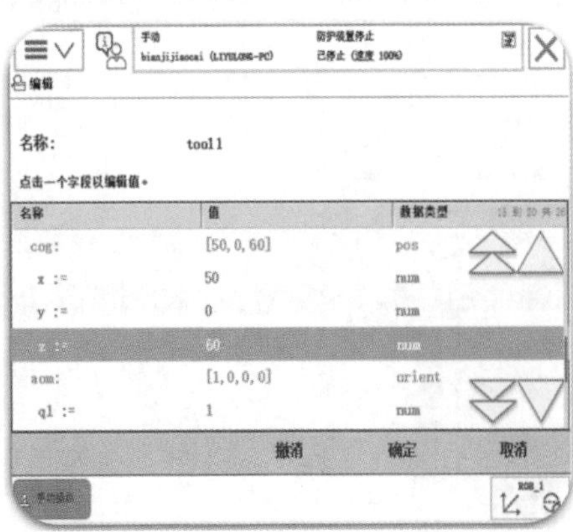

Fig. 4-32 Operation Step 20

图 4-32 操作步骤 20

（18）如图 4–33 所示，单击“确定”完成 tool1 数据更改。

(18) As shown in Fig. 4-33, click “OK” to complete the tool1 data modification.

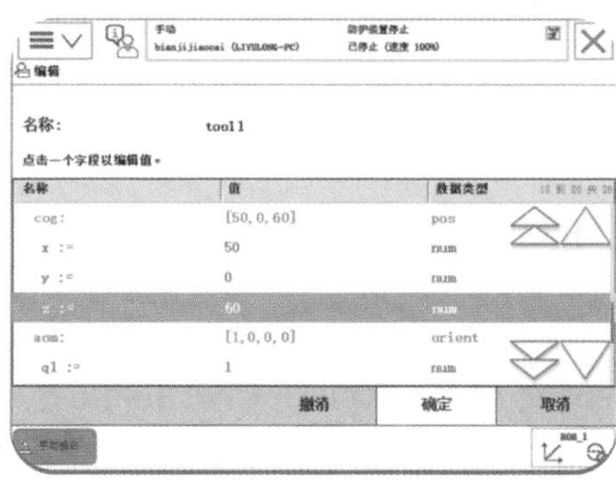

Fig.4-33 Operation Step 21

图 4–33 操作步骤 21

（19）如图 4–34 所示，按照工具重定位动作模式，把坐标系选为 “工具”；工具坐标选为 “tool1”。通过示教器操作可看见 TCP 点始终与工具参考点保持接触，而机器人根据重定位操作改变姿态。

(19) As shown in Fig.4-34, According to the tool repositioning action mode, select the coordinate system as “tool” ; select the tool coordinate as “tool1” . Through the operation of the Flexpendant, it can be seen that the TCP point is always in contact with the tool reference point, and the robot changes its posture according to the repositioning operation.

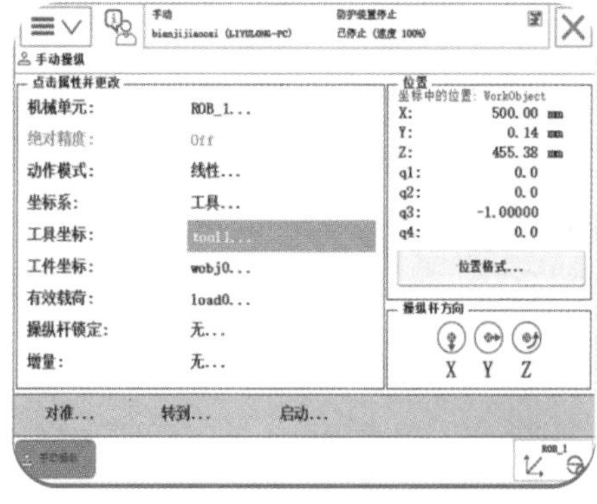

Fig.4-34 Operation Step 22

图 4–34 操作步骤 22

微信扫一扫

教学视频

Lecture 4-4
Coordinate System Setting(4)
4-4 坐标系统设置（4）

1. Setting of Workpiece Coordinate System

For the setting of the workpiece coordinate system, a workpiece coordinate system can be established by only defining three points on the plane of the object.

(1) X1 and X2 determine the positive X direction of the workpiece coordinate system;

(2) Y1 determines the positive direction of the workpiece coordinate Y;

(3) The origin of the workpiece coordinate system is the projection of Y1 on the workpiece coordinate X.

The workpiece coordinate system conforms to the right-hand rule. The right-hand rule can determine the positive and negative directions of each axis of the coordinate system and refer to the positive and negative rotation directions of each axis.

The steps to establish a workpiece coordinate system are as follows:

(1) In the manual operation panel, select “Workpiece Coordinates”. As shown in Fig. 4-35.

1. 工件坐标系的设定

工件坐标系的设定，在对象的平面上，只需定义三个点就可以建立一个工件坐标系。

（1）X1、X2 确定工件坐标系 X 正方向；

（2）Y1 确定工件坐标 Y 正方向；

（3）工件坐标系的原点是 Y1 在工件坐标 X 上的投影。

工件坐标系符合右手定则，通过右手定则可判断坐标系的各轴正负方向以及参考各轴的旋转正负方向。

建立工件坐标系的操作步骤如下：

（1） 在手动操作面板中，选择“工件坐标”。如图 4-35 所示。

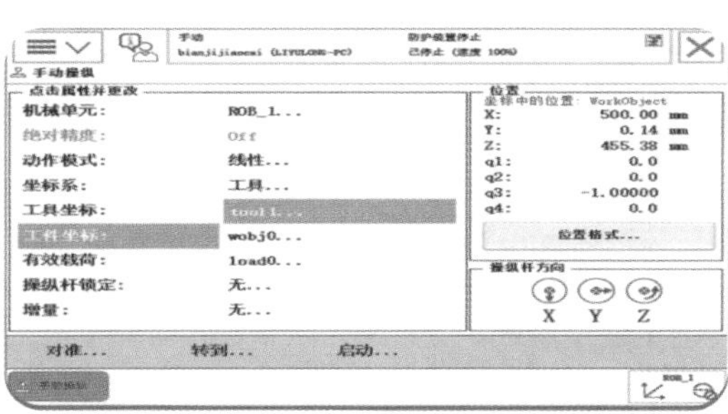

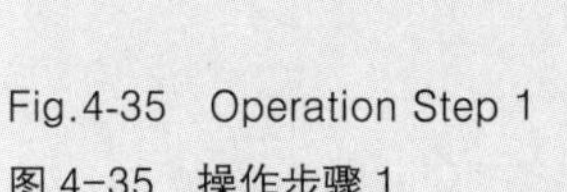

Fig.4-35　Operation Step 1

图 4-35　操作步骤 1

（2）单击“新建”。如图 4-36 所示。

(2) Click“New…”. As shown in Fig. 4-36.

Fig.4-36　Operation Step 2

图 4-36　操作步骤 2

（3）对工件数据属性进行设定后，单击“确定”。如图 4-37 所示。

(3) After setting the properties of the workpiece data, click“OK”. As shown in Fig. 4-37.

Fig.4-37　Operation Step 3

图 4-37　操作步骤 3

（4）打开编辑菜单，选择“定义”。如图 4-38 所示。

(4) Open the edit menu and select“Define”. As shown in Fig. 4-38.

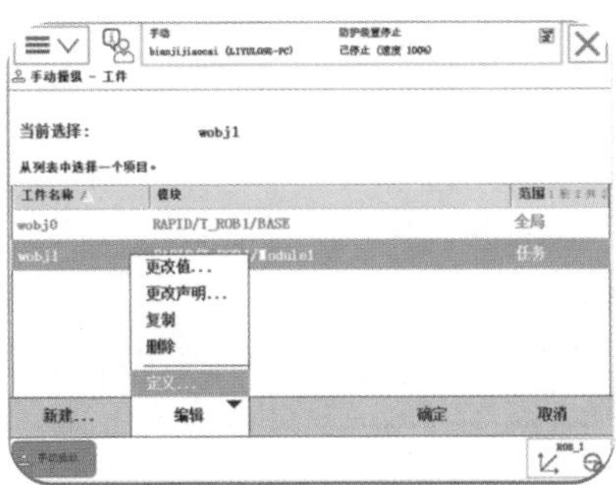

Fig.4-38 Operation Step 4

图 4-38 操作步骤 4

(5) Set the user method to "3 points". As shown in Fig. 4-39.

（5） 将用户方法设定为 "3 点"。如图 4-39 所示。

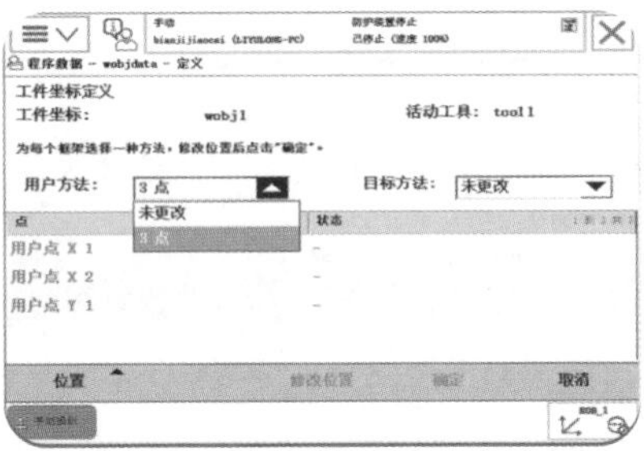

Fig.4-39 Operation Step 5

图 4-39 操作步骤 5

(6) Manually operate the tool reference point of the robot close to the X1 point of the defined workpiece coordinate system. As shown in Fig. 4-40.

（6）手动操作机器人的工具参考点靠近定义工件坐标的 X1 点。如图 4-40 所示。

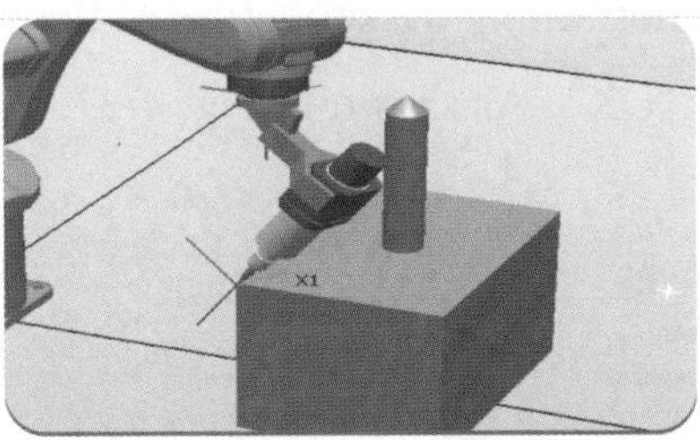

Fig.4-40 Operation Step 6

图 4-40 操作步骤 6

(7) Click "Modify Position" and record the X1 point. As shown in Fig. 4-41.

（7）单击 "修改位置"，将 X1 点记录下来。如图 4-41 所示。

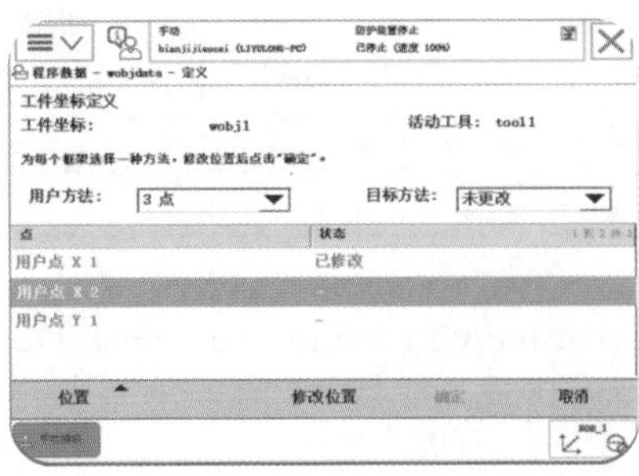

Fig.4-41 Operation Step 7

图 4-41 操作步骤 7

（8）手动操作机器人的工具参考点靠近定义工件坐标的 X2 点，然后在示教器中完成位置修改。如图 4－42 所示。

(8) Manually operate the tool reference point of the robot close to the X2 point of the defined workpiece coordinate system, and then the position modification is completed in the Flexpendant. As shown in Fig. 4-42.

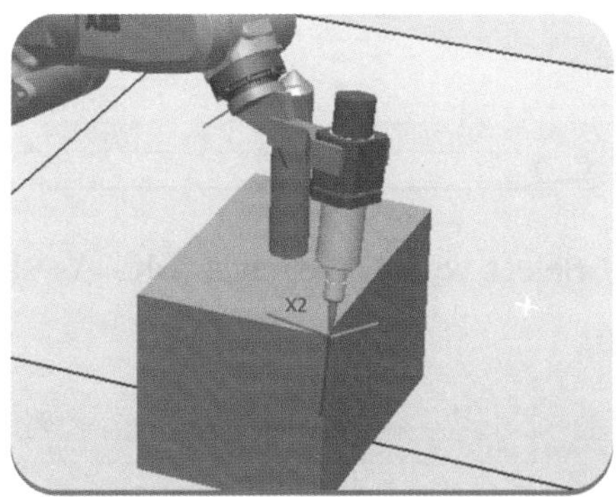

Fig.4-42　Operation Step 8
图 4－42　操作步骤 8

（9）手动操作机器人的工具参考点靠近定义工件坐标的 Y1 点，然后在示教器中完成位置修改。如图 4－43 所示。

(9) Manually operate the tool reference point of the robot close to the Y1 point of the defined workpiece coordinate system, and then the position modification is completed in the Flexpendant. As shown in Fig. 4-43.

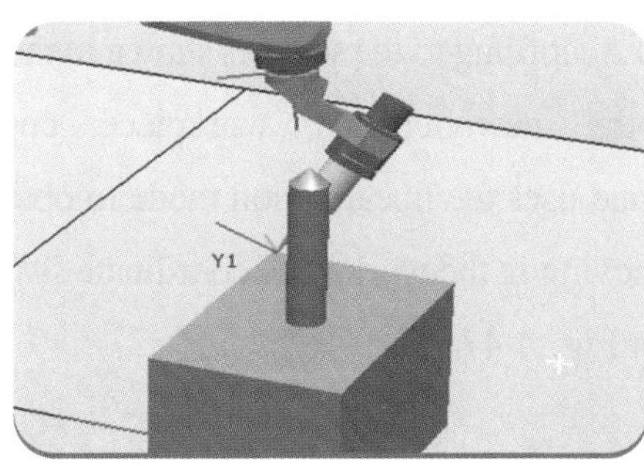

Fig.4-43　Operation Step 9
图 4－43　操作步骤 9

（10）在窗口中单击“确定”。如图 4－44 所示。

(10) Click “OK” in the window. As shown in Fig. 4-44.

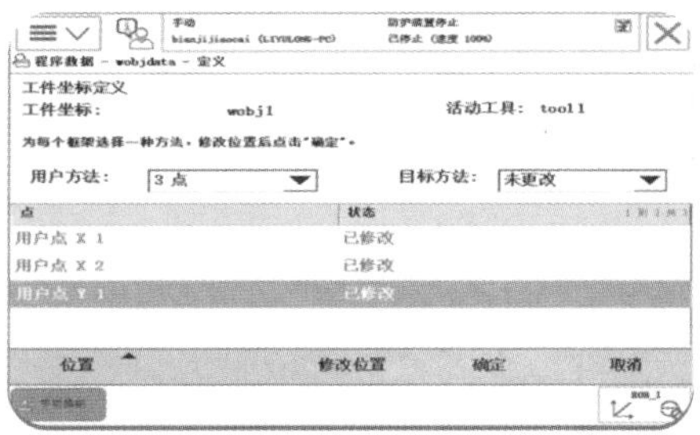

Fig.4-44　Operation Step 10
图 4－44　操作步骤 10

(11) After confirming the position of the workpiece, click “OK”. As shown in Fig. 4-45.

(11)对工件位置进行确认后，单击 “确定”。如图 4-45 所示。

Fig.4-45 Operation Step 11
图 4-45 操作步骤 11

(12) Select wobj1 and click OK. As shown in Fig. 4-46.

(12) 选中 wobj1，然后单击确定。如图 4-46 所示。

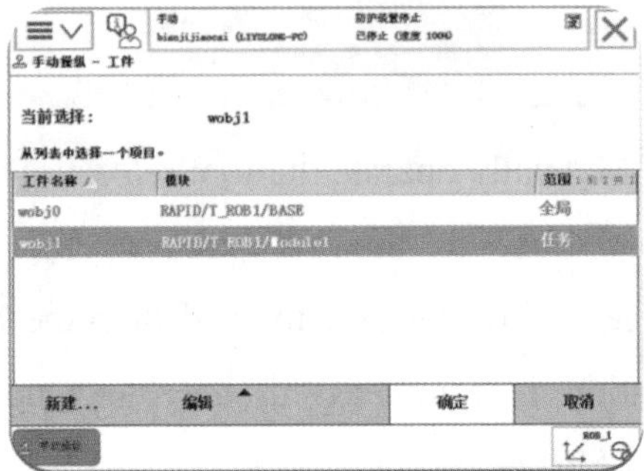

Fig.4-46 Operation Step 12
图 4-46 操作步骤 12

(13) According to the settings shown in the figure, selects the newly-created workpiece coordinate system, and uses the linear action mode to observe the way of moving in the workpiece coordinate system. As shown in Fig. 4-47.

(13) 按照如图所示的设置，坐标系选择新创建的工件坐标系，使用线性动作模式，观察在工件坐标系下移动的方式。如图 4-47 所示。

Fig.4-47 Operation Step 13
图 4-47 操作步骤 13

Lecture 5-1

Robot Programming Control(1)

5-1 机器人编程控制（1）

微信扫一扫

教学视频

工业机器人在自动模式运行时，执行存储器中的程序。ABB工业机器人的编程程序采用的是RAPID程序,接下来学习RAPID程序的概念和结构，以及程序的建立方法。

1. RAPID程序的概念

RAPID语言是一种英文编程语言，所包含的指令可以移动机器人、设置输出、读取输入，还能实现决策、重复其他指令、构造程序、与系统操作员交流等功能。不同厂家生产的机器人编程语言会有所不同，但在实现的功能上大同小异。

RAPID程序中包含一连串控制机器人的指令，执行这些指令可以实现对机器人的控制操作。RAPID程序的基本组成元素包括数据、指令和函数。

RAPID语言的结构和风格类似于C语言，例如变量名的命名规则、编程语言的语句结构还有数据类型跟C语言比较

When the industrial robot is running in automatic mode, it is executing the program in the memory. The programming of ABB industrial robot adopts the RAPID prodedure. we will learn the concept and structure of the RAPID program and the method of establishing the program.

1. Concept of RAPID Program

RAPID language is an English programming language. The instructions contained in it can move the robot, set the output, read the input, and realize the functions of decision-making, repeating other instructions, constructing programs, and communicating with system operators. Robot programming languages produced by different manufacturers will be different, but the functions implemented are similar.

The RAPID program contains a series of instructions to control the robot. Executing these instructions can realize the control operation of the robot. The basic elements of a RAPID program include data, instructions and functions.

The structure and style of the RAPID language is similar to the C language, such as the naming rules of variable names, the statement structure of the programming language and the data types are closer to the C language, so if you have the foundation of the C language, it will be easier to learn the RAPID language.

接近，因此如果有 C 语言的基础，学习 RAPID 语言会比较容易。

2. Structure of RAPID Program

2. RAPID 程序结构

The program module is used to construct the execution program of the robot. When designing the robot program,the program can create different modules according to different purposes, such as a program module for position calculation or a program module for storing data, which is convenient for classification management. The RAPID program is composed of program modules and system modules. The system modules include USER module and BASE module. The statements in the system module cannot be modified to avoid affecting the normal operation of the robot.

程序模块用于构建机器人的执行程序。在设计机器人的程序时，程序模块可以根据不同的用途创建不同的程序模块，如用于位置计算的程序模块，用于存储数据的程序模块，这样便于归类管理。RAPID 程序由程序模块和系统模块构成，系统模块里又包含了USER 模块和 BASE 模块，系统模块里的语句不能修改，以免影响机器人正常的工作。

What needs to be explained about the RAPID program is:

关于 RAPID 程序需要说明的是：

First of all,In a RAPID program, there is one and only one main program "main", and as the execution starting point of the entire RAPID program, it can exist in any program module. In the RAPID program, for function names composed of the same letters in the same order, the case of the letters does not affect the meaning of the function name. For example, both the lower case main and MAIN in capitals represent the main function, so these two function names cannot be the same function name of the routine of the RAPID program at the same time. Similarly, in a module, function names composed of the same letters in the same order cannot appear, because they represent the same routine program.

在 RAPID 程序中，有且只有一个主程序 main，并且作为整个 RAPID 程序的执行起点，可存在于任意一个程序模块中。在 RAPID 程序中，对于由相同的字母按照相同的顺序构成的函数名，字母的大小写不影响函数名的意义，例如 main 和 MAIN 都表示主函数，所以这两个函数名不能同时作为同一个RAPID程序的例行程序的函数名。同理，在一个模块中，不能出现由相同的字母按照相同的顺序构成的函数名，因为它们表示同一个例行程序。

每一个程序模块一般包含了程序数据、程序、指令和函数四种对象。程序主要分为Procedure、Function和Trap三类。Procedure相当于C语言里的子函数，Function类似于指令并且在执行完了后可以返回一个数值，Trap相当于中断处理函数。Procedure类型的程序没有返回值，可以用调用例行程序指令直接调用任何模块中的Procedure类型的程序；Function类型的程序有特定类型的返回值，必须通过表达式调用；Trap例行程序不能在程序中直接调用。

习惯上我们先创建放置主函数的模块，并给模块取一个特殊的名字，以便和其他功能模块进行区别。即可以先创建新的执行特定功能的模块，然后在模块里创建一些执行特定功能的例行程序。

3. RAPID程序的建立

在ABB工业机器人示教器中创建程序模块和例行程序的过程如下：

（1）在示教器操作界面中选择“Program Editor”选项。

（2）在模块列表界面点击左下角的“文件”菜单，然后点击“新建模块”命令。

（3）在弹出的对话框中点

Secondly, Each program module generally contains four objects: program data, program, instruction and function. The program is mainly divided into three categories: Procedure, Function and Trap. Procedure is equivalent to a sub-function in C language. Function is similar to an instruction and can return a value after execution. Trap is equivalent to an interrupted processing function. The Procedure type program has no return value. You can use the routine instruction to directly call the Procedure type program in any module; the Function type program has a specific type of return value and must be called through an expression; Trap routines cannot be directly called in the program.

It is customary that we first create a module to place the main function, and give the module a special name to distinguish it from other functional modules. That is, you can first create a new module that performs a specific function, and then create some routines that perform a specific function in the module.

3. Establishment of RAPID Program

The process of creating program modules and routines in the ABB industrial robot Flexpendant is as follows:

(1) Select the “Program Editor” option in the Flexpendant operating interface.

(2) Click the “File” menu in the lower left corner of the module list interface, and then click the “New Module” command.

(3) Click the “Yes” button in the pop-up dialog box.

(4) Enter the name of the module in the created

new module interface and select the module type. The default type of the program module is "Program", and then click the "OK" button to complete the creation of the new module. It is customary that we first create a module to place the main function, and give the module a special name to distinguish it from other functional modules.

(5) In the module list, select the newly created module, and then click the "Display Module" button.

(6) Click the "Routines" button to enter the routine creation.

(7) In the interface that displays the routine, open the "File" menu and click the "New Routine" command.

(8) First create a main program, set its name to "main", and then click the "OK" button.

(9)Other routines are created in the same way. That is, you can first create a new module that performs a specific function, and then create some routines that perform a specific function in the module.

击"是"按钮。

（4）在创建的新模块界面输入模块的名称，并选择模块类型，程序模块默认类型是"Program"，然后点击"确定"按钮完成新模块的建立。习惯上我们先创建放置主函数的模块，并给模块取一个特殊的名字，以便和其他功能模块进行区别。

（5）在模块列表中，选中模块列表中的新建模块，然后点击"显示模块"按钮。

（6）点击"Routines"按钮进入例行程序的新建。

（7）在显示例行程序的界面，打开"文件"菜单，点击"新建例行程序"命令。

（8）首先创建一个主程序，将其名称设定为"main"，然后点击"确定"按钮。

（9）其他的例行程序按照相同的办法进行创建。即可以先创建新的执行特定功能的模块，然后在模块里创建一些执行特定功能的例行程序。

Lecture 5-2
Robot Programming Control(2)
5-2 机器人编程控制（2）

微信扫一扫

教学视频

工业机器人在自动运行时，经常会做直线运动、圆弧运动，同时会设置某些数字信号，和判断信号的状态等操作。本节课我们来学习一下工业机器人的运动指令和常用 I/O 控制指令，并在 RobotStudio 软件上进行演示。

1. 运动控制指令

（1）MoveAbsJ 绝对运动指令

移动每一个关节轴运动至给定位置，常用于机器人回到机械零点的位置或 Home 点。

机器人在执行这个指令时，机器人的关节有可能会遇到奇异点。机器人到达奇异点，将会引起自由度减少，使得关节轴无法实现某些方向的运动，还有可能导致关节轴失控。一般来说，机器人有两类奇异点，分别为臂奇异点和腕奇异点。臂奇异点是指轴 4、轴 5 和轴 6 的交点与轴 1 在 Z 轴方向上的交点所处位置；腕奇异

When an industrial robot is running automatically, it often performs linear motion and circular arc motion, and at the same time it sets certain digital signals and determines the state of the signals and so on. In this lesson, we will learn about the motion instructions and common I/O control instructions of industrial robots, and demonstrate them on RobotStudio software.

1. Motion Control Instructions

(1) MoveAbsJ Absolute Motion Instruction

Move each joint axis of the robot to a given position, which is often used for the robot to return to the mechanical zero point or Home point.

When the robot executes this instruction, the joints of the robot may encounter singularities. When the robot reaches the singular point, it will cause the degree of freedom to be reduced, making the joint axis unable to move in certain directions, and may also cause the joint axis to lose control. Generally speaking, robots have two types of singularities, namely arm singularities and wrist singularities. The arm singularities refer to the location of the intersection of axis 4, axis 5, and axis 6 and the intersection of axis 1 in the Z-axis direction; the wrist singularities refer to the position of axis 4 and axis 6 on the same line (that is, the angle of axis 5 is 0) points.

Instruction Format: MoveAbsJ *\NoEoffs, v1000, z50, tool0.

Parameter description: * indicates the position data of the target point; \NoEoffs represents the external axis without offset data;

v1000 represents movement speed data, 1000mm/s;

z50 represents turning area data, which means that when the TCP point is about to reach the target point, it will avoid using an arc with a radius of 50mm. The larger the value is, the smoother the robot will move. If you want the TCP point to reach this point accurately, you can use the parameter fine;

tool0 represents tool coordinate data.

Example: To run the robot to the zero position of each joint axis, the program is as follows:

PERS jointarget

jpos10:=[[0,0,0,0,0,0], [9E+09,9E+09, 9E+09, 9E+09, 9E+09, 9E+09]];

! Each joint axis in the joint target point data is zero degree.

MoveAbsJ jpos10\NoEoffs, v1000, z50, tool0;

Here PERs is a variable, and jointarget is a data type, which represents joint data.

(2)MoveJ Joint Movement Instructions

Quickly move the TCP point of the robot to the given target point. The running track is not necessarily a straight line, the robot's motion state is

点是指轴 4 和轴 6 处于同一条线上（即轴 5 角度为 0）的点。

指令格式：MoveAbsJ *\NoEoffs，v1000，z50，tool0。

参数说明：* 表示目标点位置数据；\NoEoffs 表示外轴不带偏移数据；

v1000 表示运动速度数据 1000mm/s；

z50 表示转弯区数据，表示 TCP 点在快要到达目标点时会以半径为 50mm 的圆弧避开，该数值越大，机器人的动作越光滑和流畅，如果想让 TCP 点精确到达该点，可以使用参数 fine；

tool0 表示工具坐标数据。

例：将机器人运行至各关节轴零度位置，程序如下所示：

PERS jointarget

jpos10: = [[0,0,0,0,0,0], [9E+09, 9E+09, 9E+09, 9E+09, 9E+09, 9E+09]];

！关节目标点数据中各关节轴为零度。

MoveAbsJ jpos10\NoEoffs, v1000, z50, tool0;

这里的 PERs 是一种变量，jointarget 是数据类型，表示关节数据。

（2）MoveJ 关节运动指令

将机器人的 TCP 点快速移动至给定目标点，运行轨迹不一定是直线，机器人的运动状态不可控。该指令一般用于对机器人路径精度要求不高的场合，运动过程中，机器人运动姿态不完全可控，但运动路径保持唯一，指令适合机器人需要大范围运动时使用，不容易在运动中发生关节轴进入机械奇异点的问题。使用该指令时，需要注意的是：该指令实现两点间的移动时，两点间整个空间区域需确保无障碍物，以防止由于运动路径不可预知所造成的碰撞。

指令格式：MoveJ *, v1000, z50, tool1\\Wobj: =wobj1;

指令说明：

*表示目标位置；

v1000 表示运动速度，1000mm/s;

z50 表示转弯区数据；

tool1 表示工具坐标系 1；

Wobj 表示工件坐标；

wobj1 表示工件坐标系 1。

（3）MoveL 线性运动指令

将机器人 TCP 点沿直线运动至给定目标点，在此运动指令下，机器人运动状态可控，运动路径保持唯一，适用于对路径精度要求高的场合，如切割、涂胶、焊接等。

指令格式：MoveL *, v1000, z50, tool1\\Wobj: =wobj1;

uncontrollable. This instruction is generally used in situations where the robot path accuracy is not required to be high. During the movement, the robot motion posture is not fully controllable, but the motion path remains unique. The instruction is suitable for use when the robot needs a large range of motion, and it is not easy to cause the problem that the shaft enters the mechanical singularity in the motion. When using this instruction, you need to pay attention to: When the instruction realizes the movement between two points, the entire space between the two points must be free of obstacles to prevent collisions caused by unpredictable movement paths.

Instruction Format: MoveJ *, v1000, z50,tool1\ \Wobj:=wobj1;

Instruction description:

* Indicates the target location;

v1000 represents the movement speed, 1000mm/s;

z50 represents turning area data;

tool1 represents tool coordinate system 1;

Wobj represents the workpiece coordinate;

wobj1 represents the workpiece coordinate system 1.

(3) MoveL linear Motion Instruction

Move the TCP point of the robot to a given target point along a straight line. Under this motion instruction, the robot motion state is controllable, and the motion path remains unique. It is suitable for occasions that require high path accuracy, such as cutting, gluing, welding, etc.

Instruction Format: MoveL *, v1000, z50,tool1\

\Wobj: =wobj1;

Instruction description:

* Indicates the target location;

v1000 represents the movement speed, 1000mm/s;

z50 represents turning area data;

tool1 represents tool coordinate system 1;

Wobj represents the workpiece coordinate;

wobj1 represents the workpiece coordinate system 1.

(4)MoveC Circular motion Instruction

Move the TCP point of the robot along a circular arc to the given target point.

Instruction Format: MoveC p1, p2, v1000, z50, tool1\\Wobj: =wobj1;

Instruction description:

p1 represents the middle point of the arc, which determines the curvature of the arc;

p2 represents the end point of the arc;

v1000 represents the movement speed, 1000mm/s;

z50 represents turning area data;

tool1 represents tool coordinate system 1;

Wobj represents the workpiece coordinate;

wobj1 represents the workpiece coordinate system 1.

If you want to use this instruction to draw a circle, you need to call the MoveC instruction twice to draw two semicircles respectively.

2. Common I/O Control Instructions

(1) Set Instruction

Set the digital output signal to 1.

指令说明：

*表示目标位置；

v1000 表示运动速度，1000mm/s；

z50 表示转弯区数据；

tool1 表示工具坐标系 1；

Wobj 表示工件坐标；

wobj1 表示工件坐标系 1。

（4）MoveC 圆弧运动指令

将机器人 TCP 点沿圆弧运动至给定目标点。

指令格式：MoveC p1, p2, v1000, z50, tool1\\Wobj: =wobj1;

指令说明：

p1 表示圆弧中间点，它决定了圆弧曲率；

p2 表示圆弧终点；

v1000 表示运动速度，1000mm/s；

z50 表示转弯区数据；

tool1 表示工具坐标系 1；

Wobj 表示工件坐标；

wobj1 表示工件坐标系 1。

如果要用这个指令画一个圆，需要调用两次这个 MoveC 指令，分别画两个半圆。

2. 常用 I/O 控制指令

（1）Set 指令

将数字输出信号置为 1。

例如：Set Do1;

（2）Reset 指令

将数字输出信号置为 0。

例如：Reset Do1；

（3）WaitDI 指令

等待一个数字输入信号变为指定状态。

例如：WaitDI Di1，1；

3. 指令的使用演示

（1）打开 RobotStudio 软件，点击【文件】—【新建】—【空工作站解决方案】—【解决方案名称】可以用默认名称，位置路径必须是全英文的形式，最后点击【创建】。

（2）点击【ABB 模型库】选择 IRB120，点击【机器人系统】，选择【从布局】，选择电脑中安装的 RobotWare 版本，这里选择 6.07.01.00 版。

（3）等到软件右下角的机器人状态变为绿色时。点击【控制器】—【示教器】—【虚拟示教器】，进入虚拟示教器后，选择【Program Editor】，进入程序模块 Module1。

（4）将示教器上的运行模式设置为手动运行。这时在 Module1 里面点击【Add Instruction】，添加 MoveAbsJ 指令。

（5）双击*，进入关节位置界面。单击【New】，进入“New-DataDeclaration”界面，名字若不用变，可点击【OK】。软件自动回到上一个界面，然后

For example: Set Do1;

(2) Reset Instruction

Set the digital output signal to 0.

For example: Reset Do1;

(3) WaitDI Instruction

Wait for a digital input signal to become the specified state.

For example: WaitDI Di1,1;

3. Demonstration of the Use of Instructions

(1) Open RobotStudio software, click [File]–[New] – [Empty Workstation Solution]

[Solution Name] can use the default name, the location path must be in English, and finally click [Create].

(2)Click [ABB Model Library] to select IRB120, click [Robot System], select [From Layout...], select RobotWare version installed on the computer, here select 6.07.01.00 version.

(3) When the robot status in the lower right corner of the software turns green, Click [Controller] –[Flexpendant]–[Virtual Flexpendant], after entering the virtual Flexpendant, select [Program Editor] to enter the program module Module1.

(4)Set the operating mode on the Flexpendant to manual operation. At this time, click [Add Instruction] in Module1 to add the MoveAbsJ instruction.

(5) Double-click * to enter the joint position interface. Click [New] to enter the“NewDataDeclaration” interface. If the name doesn't need to be changed, click [OK]. The software automatically returns to the previous interface, and then click [OK]. Similarly, add MoveJ instruction, MoveL instruction and

MoveC instruction in sequence.

点击【OK】。同理依次添加MoveJ 指令、MoveL 指令和MoveC 指令。

教学视频

Lecture 5-3
Robot Programming Control(3)
5-3 机器人编程控制（3）

工业机器人在自动运行时，经常会执行一些逻辑判断、循环动作和延时等待指令。本节课学习工业机器人的逻辑判断指令、循环指令和延时指令。

Compact IF 指令。紧凑型条件判断指令，用于当条件满足时，执行对应的程序。

例如：IF reg1>5 THEN
Set do1;
ENDIF

如果条件 reg1>5 满足，则执行 Set do1 指令。

IF 指令。条件判断指令，满足 IF 条件，则执行满足该条件下的指令。

例如：IF reg1>5 THEN
Set do1;
Set do2;
ENDIF

仅当 reg1 大于 5 时，设置信号 do1 和 do2。

例如：IF counter>100 THEN
counter: =100;

When an industrial robot is running automatically, it often executes some instructions such as logical judgments, loop and delays waiting. In this lesson, we are going to learn the logical judgment instructions, loop instructions and delay instructions of industrial robots.

Compact IF instruction. The compact condition judgment instruction is used to execute the corresponding program when the condition is met.

E.g: IF reg1>5 THEN
Set do1;
ENDIF

If the condition reg1>5 is met, the Set do1 instruction is executed.

IF instruction. Condition judgment instruction, if the IF condition is met, the instruction under the condition is executed.

E.g: IF reg1>5 THEN
Set do1;
Set do2;
ENDIF

Only when reg1 is greater than 5, the signals do1 and do2 are set.

E.g: IF counter>100 THEN

```
        counter:=100;
    ELSEIF counter<0 THEN
        counter:=0;
    ELSE
        counter:=counter+1;
    ENDIF
```

ELSE must be paired with IF. The principle of pairing is to pair with the nearest IF before the ELSE, so the ELSE and ELSEIF in the program are paired. In the program, if the counter is between 0 and 100, the counter incremented by 1, and if the counter exceeds this range, the corresponding limit is assigned to the counter.

TEST instruction. According to the judgment result of the formulated variable, the corresponding program is executed.

E.g: TEST reg1

```
CASE 1:
rountine1;
CASE 2:
rountine2;
DEFAULT:
Stop;
ENDTEST
```

Determine the value of reg1, if it is 1, execute rountine1; if it is 2, execute rountine2, otherwise execute stop.

Loop FOR instruction. Repeated execution of judgment instructions is used when one or more instructions need to be executed multiple times.

E.g:

FOR i FROM 1 TO 10 DO

```
ELSEIF counter<0THEN
counter: =0;
ELSE
counter: =counter+1;
ENDIF
```

ELSE 必须和 IF 配对使用，配对的原则是和 ELSE 之前最近的没有 ELSE 配对的 IF 进行配对，所以程序中的 ELSE 和 ELSEIF 进行配对。程序中 counter 如果在 0～100 之间，则 counter 进行自增 1；如果 counter 超过了这个范围，则向 counter 分配相应的限值。

TEST 指令。根据制定变量的判断结果，执行对应程序。

例如：TEST reg1

```
CASE 1:
rountine1;
CASE 2:
rountine2;
DEFAULT:
Stop;
ENDTEST
```

判断 reg1 数值，若为 1 则执行 rountine1；若为 2 则执行 rountine2，否则执行 stop。

循环 FOR 指令。重复执行判断指令，用于一个或多个指令需要重复执行多次的情况。

例如：

FOR i FROM 1 TO 10 DO

routine1;

ENDFOR

重复执行 10 次 routine1 里的程序。

WHILE 指令。条件判断指令，用于满足给定条件的情况下，重复执行对应程序。

例如：

WHILE reg 1< reg2 DO

reg1: =reg1+1;

ENDWHILE

如果变量 reg1<reg2 条件一直成立，则重复执行 reg1 加 1，直至 reg1<reg2 条件不成立为止。

紧凑型条件判断指令 IF，是只有满足条件时才能执行指令；条件判断指令 IF，基于是否满足条件，执行指令序列；重复执行判断指令 FOR，重复一段程序多次，可以简化程序语句；条件判断指令 WHILE，重复指令序列，直到不满足给定条件。TEST 指令可以对不同情况进行处理。

WaitTime 时间等待指令，它用于程序中等待一个指定的时间，再往下执行程序。该指令经常用在气爪抓取物品或吸盘吸放物品时，物品能可靠地被抓取或吸取的场合。

例如：WaitTime 3;

指令表示等待 3s 以后，程序向下执行。

routine1;

ENDFOR

Repeat the program in routine1 for 10 times.

WHILE instruction. The condition judgment instruction is used to repeatedly execute the corresponding program when the given condition is met.

E.g:

WHILE reg 1< reg2 DO

reg1:=reg1+1;

ENDWHILE

If the condition of variable reg1<reg2 is always established, repeat reg1 plus 1 until the condition of reg1<reg2 is not established.

The compact conditional judgment instruction IF can only execute the instruction when the condition is met; the conditional judgment instruction IF executes the sequence of instructions based on whether the condition is met; the repeated execution of the judgment instruction FOR, repeats a program many times, can simplify the program statement; the conditional judgment instruction WHILE, repeat the instruction sequence until the given conditions are not met. The TEST instruction can handle different situations.

WaitTime time-waiting instruction,which is used in the program to wait for a specified time before executing the program. This instruction is often used when the air claw grabs the object or the suction cup sucks and releases the object, and the object can be grasped or sucked reliably.

E.g: WaitTime 3;

The instruction indicates that after waiting 3s, the program executes downward.

It should be noted that if the Move instruction is used before the instruction, the Move instruction must be programmed through the stop point (fine) instead of the flyover point (that is, Z is a point with a value). Otherwise, it is impossible to restart after a power failure.

Offs offset function instruction. This instruction takes the selected target point as the reference and offsets a certain distance along the X, Y, Z axis of the selected workpiece coordinate system. This function has a return value, that is, the result of calling this function is a value of a certain data type. It cannot be used as a single line of statement when used. It needs to be called by assignment or as a variable of other functions.

E.g:

MoveLOffs (p10,0,0,10), v1000, z50, tool0\Wobj:=wobj1;

Move the robot TCP to a position that is offset by 10 mm along the positive Z axis of wobj1 with p10 as the reference point.

p10:=offs(p10,0,0,10);

Using p10 as the benchmark, assign p10 to p10 at a position offset by 10 mm in the positive direction of the Z axis.

RelTool offset instruction. This instruction is also an offset instruction, and the angle offset can be set, but the reference coordinate system is the tool coordinate system, such as:

MoveL RelTool(p10,0,0,10\Rx:=20\Ry:=30\Rz:=40),v200,z10,tool1;

Then the robot TCP moves to the reference point

需要注意的是，如果在指令之前采用 Move 指令，则必须通过停止点（fine）而非飞越点（即 Z 是有数值的点）来编程 Move 指令。否则，不可能在电源故障后重启。

Offs 偏移功能指令。该指令以选定的目标点为基准，沿着选定工件坐标系的 X、Y、Z 轴方向偏移一定的距离。该函数是有返回值的，即调用此函数的结果是得到某一数据类型的值，在使用时不能单独作为一行语句，需要通过赋值或者作为其他函数的变量来调用。

例如：

MoveL Offs(p10,0,0,10), v1000, z50, tool0\Wobj: = wobj1;

将机器人 TCP 移动至以 p10 为基准点，沿着 wobj1 的 Z 轴正方向偏移 10mm 的位置。

p10: =offs（p10,0,0,10）;

以 p10 为基准，将 p10 在 Z 轴正方向偏移 10mm 的位置赋值给 p10。

RelTool 偏移指令。这个指令同样为偏移指令，而且可以设置角度偏移，但其参考的坐标系为工具坐标系，如：

MoveL RelTool（p10,0,0,10\Rx:=20\Ry:=30\Rz:=40），v200, z10, tool1;

机器人 TCP 移动至以 p10 为基准点，沿 tool1 坐标系 Z 轴正方向偏移 10mm，且 TCP 绕着 tool1 坐标系 X 轴旋转 20°，绕着 Y 轴旋转 30°，绕着 Z 轴旋转 40 度。

在 RobotStudio 软件里进行指令的操作。第一步，按照之前的办法创建机器人工作站，并打开虚拟示教器的程序编辑器，进入程序模块 Module1；第二步，将机器人的运动模式设置为手动；第三步，在程序模块 Module1 中点击【Add Instruction】编写程序；第四步，点击<EXP>进入编辑界面，点击【New】，进入定义变量的界面，变量名为 bPickOK，【Scope】【Storage type】选择默认；第五步，在插入表达式界面，点击【Edit】【All】，输入表达式 bPickOK=False，点击【OK】；第六步，点击<SMT>，点击【Add Instruction】，添加其他指令。

p10, offset 10mm along the positive Z axis of the tool1 coordinate system, and the TCP rotates 20° around the X axis of the tool1 coordinate system, 30° around the Y axis, and 40°around the Z axis.

Operating instruction in robotstudio software. The first step is to create a robot workstation according to the previous method, and open the program editor of the virtual Flexpendant, enter the program module Module1; The second step, set the robot's motion mode to manual; The third step, click [Add Instruction] in the program Module1 to write the program; The fourth step, click <EXP> to enter the editing interface, click [New] to enter the interface for defining variables, the variable name is bPickOK, select the default option of [Scope] and [Storage type]. The fifth step, in the insert expression interface, click [Edit] [All], enter the expression bPickOK=False, and click [OK]; The sixth step, click <SMT>, click [Add Instruction], add other instructions.

微信扫一扫

教学视频

Lecture 5-4

Robot Programming Control(4)

5-4 机器人编程控制（4）

When writing a robot program, some variables are often assigned. In this lesson, we are going to learn the assignment instructions, commonly-used mathematical operation instructions, and program data storage types of ABB industrial robots, and demonstrate them on RobotStudio software.

We need to define variables or constants and need to assign values to them. Assignment instructions are used to assign values to the data in the program. The method of assignment can be to assign a constant to the program data, or to assign a mathematical expression to the program data. E.g:

① Constant assignment: reg1:=5.

② Expression assignment: reg2:=reg1+4.

When assigning data, the variable and value data types must be the same. When the program is running, constant data is not allowed to be assigned.

Commonly-used mathematical operations instructions. The first is the Clear instruction. It is used to clear numeric variables or permanent data objects, that is, to set the value to 0.

E.g: Clear reg1;

Reg1 is cleared, that is, reg1:=0.

The second is the ADD instruction. It is used to

在编写机器人程序时，经常会给一些变量进行赋值操作。本节学习ABB工业机器人的赋值指令、常用的数学运算指令、程序数据存储类型，并在RobotStudio软件上进行操作。

要定义变量或常量，需要给其赋值。赋值指令用于对程序中的数据进行赋值，赋值的方式可以为将一个常量赋值给程序数据，还可以为将数学表达式赋值给程序数据。例如：

①常量赋值：reg1：=5。

②表达式赋值：reg2：=reg1+4。

数据赋值时，变量与值数据类型必须相同。程序运行时，常量数据不允许赋值。

常用的数学运算指令。第一个是Clear清除指令。它用于清除数值变量或永久数据对象，即将数值设置为0。

例如：Clear reg1;

Reg1得以清除，即，reg1：=0。

第二个是 ADD 指令。它用于从数值变量或者永久数据对象增减一个数值。

例如：Add reg1,3;

将 Reg1 增加 3，即，reg1:=reg1+3。

Add reg1，−reg2;

将 Reg2 从 reg1 中减去，即，reg1: =reg1−reg2。

第三个是 Incr 增一指令。它用于向数值变量或者永久数据对象增加 1。

例如：WHILE reg 1<101 DO

Sum: =Sum+reg1;

Incr reg1;

ENDWHILE

第四个是 Decr 减一指令。它用于从数值变量或者永久数据对象减去 1，与 Incr 用法一样，但是作用刚好相反。

程序数据的存储类型。程序数据的存储类型包括变量 VAR、可变量 PERS 以及常量 CONST。具体来说，首先是变量 VAR。变量进行定义时，可以赋值，也可以不赋值。在程序中遇到新的赋值语句，当前值改变，但初始值不变，遇到指针重置（指针重置是指程序指针被人为地从一个例行程序移至另一个例行程序或者程序指针移至 main）又恢复到初始

add or subtract a value from a numeric variable or permanent data object.

E.g: Add reg1,3;

Increase Reg1 by 3, that is, reg1:=reg1+3.

Add reg1,−reg2;

Subtract Reg2 from reg1, that is, reg1:=reg1−reg2.

The third one is the Incr increment instruction. It is used to add 1 to a numeric variable or permanent data object.

E.g: WHILE reg 1<101 DO

Sum:=Sum+reg1;

Incr reg1;

ENDWHILE

The fourth is Decr minus one instruction. It is used to subtract 1 from a numeric variable or a permanent data object, which is the same as Incr, but has the opposite effect.

The storage type of program data. The storage types of program data include variable VAR, variable PERS and constant CONST. Specifically, the first is the variable VAR. When a variable is defined, it can be assigned or not assigned. When a new assignment statement is encountered in the program, the current value changes, but the initial value does not change. Encountered pointer reset (pointer reset refers to the program pointer is artificially moved from one routine to another or the program pointer moved to main) and restored to its initial value. Variables can be introduced through variable declarations, as well as global system, task layout or local variable declarations. Examples are as follows:

VAR num length:=0;

Assign 0 to the data variable length,

VAR string author_name:= “John Smith”;

Assign the string “John Smith” to the string variable author_name.

TASK VAR num taskvar:= 456;

Assign 456 to the variable taskvar in the routine.

LOCAL VAR num localvar:=789;

Assign 789 to the variable localvar in the current module.

The second is variable PERS. Regardless of the program’s pointer, it will retain the last assigned value. When defining, all variables must be assigned a corresponding initial value, and the initial value of variables cannot be assigned using expressions. Examples are as follows:

PERS num nbr:=1;

Assign 1 to the data variable nbr.

PERS string text:= “Hello”;

Assign the character “Hello” to the string variable text.

Finally, the constant CONST. It is assigned a specific value when it is defined, and cannot be changed in the program, but can only be modified manually. When defining, all constants must be assigned a corresponding initial value. Examples are as follows:

CONST num gravity:=9.8;

Assign 9.8 to the data type constant gravity.

CONST string greating:= “Hello”;

Assign the string “Hello” to the string constant greating.

值。可通过变量声明引入变量，同时也可作系统全局、任务布局或局部变量声明。示例如下：

VAR num length: =0;

把0赋值给数据变量length，

VAR string author_name : = “John Smith”;

把“John Smith”这个字符串赋值给字符串变量author_name。

TASK VAR num taskvar : = 456;

把456赋值给例行程序中的变量taskvar。

LOCAL VAR num localvar : =789;

把789赋值给当前模块中的变量localvar。

其次是可变量PERS。不管程序的指针如何，它都会保持最后被赋予的值。在定义时，所有可变量必须被赋予一个相应的初始值，可变量的初始值不能使用表达式赋值。示例如下：

PERS num nbr: =1;

把1赋值给数据可变量nbr。

PERS string text:= “Hello”;

把字符“Hello”赋值给字符串可变量text。

最后是常量CONST。它在定义时就被赋予了特定的数值，并不能在程序中进行改动，只能手动进行修改。在定义时，所有常量必须被赋予一个相应的初始值。示例如下：

CONST num gravity:=9.8;

把 9.8 赋值给数据型常量 gravity。

CONST string greating: = “Hello”;

把字符串” Hello” 赋值给字符串常量 greating。

接下来在 RobotStudio 软件里对常量和变量的定义进行操作，步骤如下：

（1）将机器人的运动模式设置为手动。

（2）按照之前的办法创建机器人工作站，选择主界面里的【Program Data】。

（3）在右下角选择【View】，选择【All Data Types】。

（4）选择需要定义的数据的类型，并进行设置。在这里需要注意，常量和变量的名字不能是关键字，要符合命名规则。第一，数据名称只能使用字母、数字及下划线；第二，数据名称的长度最多不能超过 16 个字符；第三，首字母必须是英文字母。Scope 有三个选项，分别是 Global、Local 和 Task。Global 表示所有模块与例行程序都可以调用的数据；Local 表示只有当前模块及其包含的例行程序才能调用的数

Next, operate the definitions of constants and variables in the RobotStudio software The steps are as follows:

(1)Set the motion mode of the robot to manual.

(2)Create a robot workstation according to the previous method, select [Program Data] in the main interface.

(3)Select [View] in the lower right corner, select [All Data Types].

(4)Select the type of data to be defined and set it. It should be noted here that the names of constants and variables cannot be keywords, but must conform to the naming rules. First, the data name can only use letters, numbers and underscores; second, the length of the data name cannot exceed 16 characters at most; third, the first letter must be an English letter. Scope has three options, namely Global, Local and Task. Global means data that can be called by all modules and routines; Local means data that can be called only by the current module and the routines it contains; Task means that it is defined in the routine, and only the routine that defines the data can be called. When the program executes to this routine, the data is called. After the current routine is executed, the data value is lost. Each time this routine is executed, the data will be the initial value.

据；Task 表示被定义在例行程序内，也只在定义该数据的例行程序才可以调用。当程序执行到此例行程序时，该数据才被调用，执行完当前此例行程序之后，该数据值丢失。每执行此例行程序时，该数据都将为初始值。

教学视频

Lecture 6-1

Parameter Setting and Project Management of Robot System(1)

6-1 参数设置和机器人系统的项目管理（1）

1.ABB工业机器人的通讯

ABB机器人提供了丰富I/O通信接口，如ABB的标准通信，与PLC的现场总线通信，还有与PC机的数据通信等都可以轻松地实现与周边设备的通信。ABB的标准I/O板提供的常用信号处理有数字量输入，数字量输出，组输入，组输出，模拟量输入，模拟量输出。ABB机器人可以选配标准ABB的PLC，省去了原来与外部PLC进行通信设置的麻烦，并且在机器人的示教器上就能实现与PLC的相关操作。本节将以常用的ABB标准I/O板DSQC651、DSQC652、DSQC653、DSQC355A、DSQC377A为例，对相关参数定义进行详细的讲解。

2. ABB 标准 I/O 板 DSQC 651

DSQC651板主要提供8个

1. Communication of ABB Industrial Robots

ABB robots provide a wealth of I/O communication interfaces, such as ABB's standard communication, fieldbus communication with PLC, and data communication with PCs, which can easily communicate with peripheral devices. The common signal processing provided by ABB's standard I/O board includes digital input, digital output, group input, group output, analog input and analog output. ABB robots can be equipped with standard ABB PLC, which saves the trouble of communicating with external PLC, and can realize PLC-related operations on the robot's Flexpendant. In this lesson, we will take the commonly-used ABB standard I/O boards DSQC651, DSQC652, DSQC653, DSQC355A and DSQC377A as examples to explain the definitions of related parameters in detail.

2. ABB Standard I/O Board DSQC651

The DSQC651 board mainly provides processing of 8 digital input signals, 8 digital output signals and 2 analog output signals. The description of the module interface is shown in the Fig. 6-1 below.

数字输入信号、8 个数字输出信号和 2 个模拟输出信号的处理。模块接口说明如图 6-1 所示。

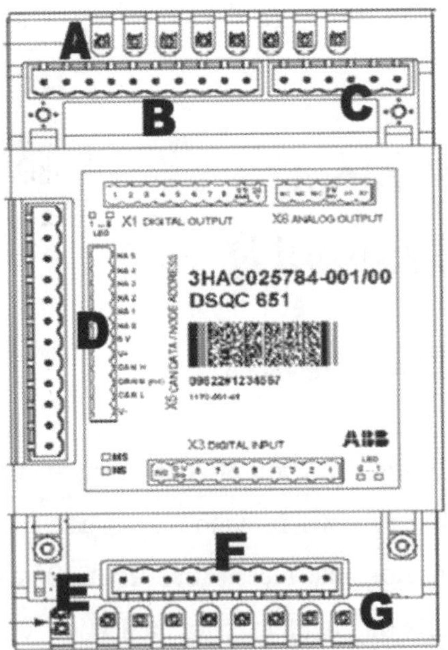

Fig.6-1 Module interface description

图 6-1 模块接口说明

In the figure, area A is the digital output signal indicator, area B is the X1 digital output interface, area C is the X6 analog output interface, area D is the X5 DeviceNet interface, area E is the module status indicator, and area F is the X3 digital input Interface, area G is the digital input signal indicator. Among them, the range of analog output is 0 ~ 10V.

图中，A 区为数字输出信号指示灯，B 区为 X1 数字输出接口，C 区为 X6 模拟输出接口，D 区为 X5 DeviceNet 接口，E 区为模块状态指示灯，F 区为 X3 数字输入接口，G 区为数字输入信号指示灯。其中，模拟输出的范围为 0 ~ 10V。

The address allocation of the digital output signal X1 terminal is shown in the Tab. 6-1 below.

数字输出信号 X1 端子的地址分配如表 6-1 所示。

Tab. 6-1 Address assignment of X1 terminal

表 6-1 X1 端子的地址

X1 Terminal No.	Use Definition	Address Allocation
1	OUTPUT CH1	32
2	OUTPUT CH2	33
3	OUTPUT CH3	34
4	OUTPUT CH4	35
5	OUTPUT CH5	36
6	OUTPUT CH6	37
7	OUTPUT CH7	38
8	OUTPUT CH8	39
9	0V	
10	24V	

数字输入信号 X3 端子的地址分配如表 6-2 所示。

The address allocation of the digital input signal X3 terminal is shown in the Tab. 6-2 below.

Tab. 6-2 Address assignment of X3 terminal
表 6-2 X3 端子的地址

X3 Terminal No.	Use Definition	Address Allocation
1	INPUT CH1	0
2	INPUT CH2	1
3	INPUT CH3	2
4	INPUT CH4	3
5	INPUT CH5	4
6	INPUT CH6	5
7	INPUT CH7	6
8	INPUT CH8	7
9	0V	
10	Unused	

DeviceNet 接口 X5 端子的连接如表 6-3 所示。

The connection of the X5 terminal of the DeviceNet interface is shown in the following Tab. 6-3.

Tab. 6-3 Address assignment of X5 terminal
表 6-3 X5 端子的地址

X5 Terminal No.	Use Definition
1	0V BLACK (black)
2	CAN signal line low BLUE (blue)
3	Shielded wire
4	CAN signal line high WHITE (white)
5	24V RED (red)
6	common terminal of GND address selection
7	Module ID bit 0 (LSB)
8	Module ID bit 1 (LSB)
9	Module ID bit 2 (LSB
10	Module ID bit 3 (LSB)
11	Module ID bit 4 (LSB)
12	Module ID bit 5 (LSB)

模拟输出接口 X6 端子接线如表 6-4 所示。

The wiring of the X6 terminal of the analog output interface is shown in the following Tab. 6-4.

Tab. 6-4 Address assignment of X6 terminal
表 6-4 X6 端子的地址

X6 Terminal No.	Use Definition	Address Allocation
1	Unused	
2	Unused	
3	Unused	
4	0V	
5	Analog output AO1	0-15
6	Analog output AO2	16-31

As the ABB standard IO board is hung on the DeviceNet network, the address of the module in the network must be set. The wire jumper 6～12 of terminal X5 is used to determine the address of the module, and the available address range is 10～63. As shown in the Fig. 6-2, cut off the wire jumpers on the 8th and 10th pins, and 2+8=10 can get the address of 10.

由于 ABB 标准 IO 板是挂在 DeviceNet 网络上的，所以要设定模块在网络中的地址。端子 X5 的 6～12 的跳线就是用来决定模块的地址的，地址可用范围为 10～63。如图 6-2 所示，将第 8 脚和第 10 脚的跳线剪去，2+8=10 就可以获得 10 的地址。

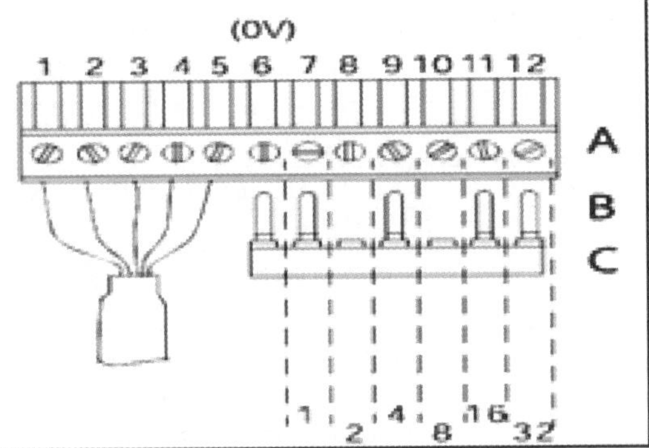

Fig.6-2 Address available range
图 6-2 地址可用范围

3. ABB Standard I/O Board DSQC652

The DSQC652 board mainly provides processing of 16 digital input signals and 16 digital output signals. The module interface description is shown in the Fig. 6-3.

3. ABB 标准 I/O 板 DSQC 652

DSQC652 板主要提供 16 个数字输入信号和 16 个数字输出信号的处理。模块接口说明如图 6-3 所示。

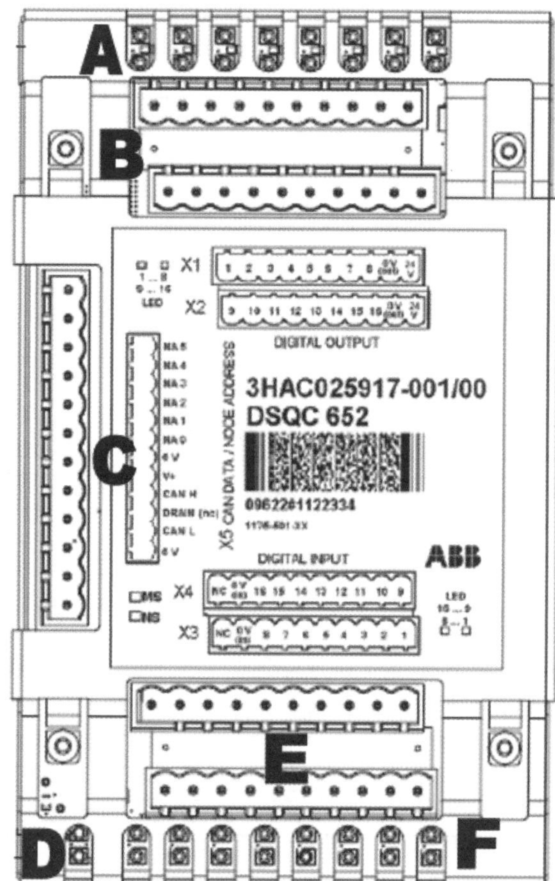

Fig.6-3 The module interface description

图 6-3 模块接口说明

其中A区为数字输出信号指示灯，B 区为数字输出接口 X1 和 X2, C 区为 DeviceNet 接口 X5, D 区为模块状态指示灯，E 区为数字输入接口 X3 和 X4，F 区为数字输入信号指示灯。

其中数字输出接口 X1 和 X2 端子的接口说明如表 6-5 所示。

Area A is the digital output signal indicator, area B is the digital output interface X1 and X2, area C is the DeviceNet interface X5, area D is the module status indicator, area E is the digital input interface X3 and X4, and area F is the digital input signal indicator.

The interface description of the digital output interface X1 and X2 terminals is shown in the following Tab. 6-5.

X1 Terminal No.	Use Definition	Address Allocation
1	OUTPUT CH1	0
2	OUTPUT CH2	1
3	OUTPUT CH3	2
4	OUTPUT CH4	3
5	OUTPUT CH5	4
6	OUTPUT CH6	5
7	OUTPUT CH7	6
8	OUTPUT CH8	7
9	0V	
10	24V	

X2 Terminal No.	Use Definition	Address Allocation
1	OUTPUT CH9	8
2	OUTPUT CH10	9
3	OUTPUT CH11	10
4	OUTPUT CH12	11
5	OUTPUT CH13	12
6	OUTPUT CH14	13
7	OUTPUT CH15	14
8	OUTPUT CH16	15
9	0V	
10	24V	

Tab. 6-5 interface X1 and X2 terminals

表 6-5 X1 和 X2 端子接口说明

The digital input interface X4 interface description is shown in the following Tab. 6-6.

数字输入接口 X4 接口说明如表 6-6 所示。

X4 Terminal No.	Use Definition	Address Allocation
1	INPUT CH9	8
2	INPUT CH10	9
3	INPUT CH11	10
4	INPUT CH12	11
5	INPUT CH13	12
6	INPUT CH14	13
7	INPUT CH15	14
8	INPUT CH16	15
9	0V	
10	Unused	

Tab. 6-6 interface X4 description

表 6-6 接口 X4 接口说明

4. ABB 标准 I/O 板 DSQC 653

DSQC653 板主要提供 8 个数字输入信号和 8 个数字继电器输出信号的处理。模块接口说明如下图 6-4 所示。

4. ABB Standard I/O Board DSQC653

The DSQC653 board mainly provides processing of 8 digital input signals and 8 digital relay output signals. The module interface description is shown in the Fig. 6-4.

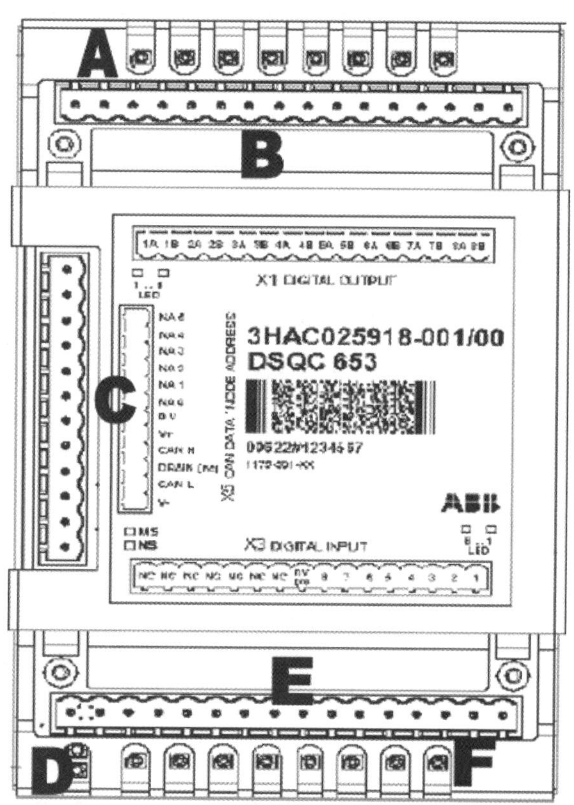

Fig.6-4 The module interface description

图 6-4 模块接口说明

其中 A 区代表数字继电器输出信号指示灯，B 区代表数字继电器输出信号接口 X1，C 区代表 DeviceNet 接口 X5，D 区代表模块状态指示灯，E 区代表数字输入信号接口 X3，F 区代表数字输入信号指示灯。

其中，X1 端子接口说明如表 6-7 所示。

Area A represents digital relay output signal indicator, area B represents digital relay output signal interface X1, area C represents DeviceNet interface X5, area D represents module status indicator, area E represents digital input signal interface X3, and area F represents digital input signal indicator.

Among them, the description of X1 terminal interface is shown in the following Tab. 6-7.

Tab. 6-7 interface X1 description

表 6-7 接口 X1 说明

X1 Terminal No.	Use Definition	Address Allocation
1	OUTPUT CH1A	0
2	OUTPUT CH1B	
3	OUTPUT CH2A	1
4	OUTPUT CH2B	
5	OUTPUT CH3A	2
6	OUTPUT CH3B	
7	OUTPUT CH4A	3
8	OUTPUT CH4B	
9	OUTPUT CH5A	4
10	OUTPUT CH5B	
11	OUTPUT CH6A	5
12	OUTPUT CH6B	
13	OUTPUT CH7A	6
14	OUTPUT CH7B	
15	OUTPUT CH8A	7
16	OUTPUT CH8B	

The digital input signal X3 interface description is shown in the following Tab. 6-8.

数字输入信号 X3 接口说明如表 6-8 所示。

Tab. 6-8 interface X3 description

表 6-8 接口 X3 说明

X3 Terminal No.	Use Definition	Address Allocation
1	INPUT CH1	0
2	INPUT CH2	1
3	INPUT CH3	2
4	INPUT CH4	3
5	INPUT CH5	4
6	INPUT CH6	5
7	INPUT CH7	6
8	INPUT CH8	7
9	0V	
10-16	Unused	

5. ABB Standard I/O Board DSQC355A

The DSQC355A board mainly provides processing of 4 analog input signals and 4 analog output signals. The module interface description is shown in the Fig. 6-5.

5. ABB 标准 I/O 板 DSQC 355A

DSQC355A 板主要提供 4 个模拟输入信号和 4 个模拟输出信号的处理。模块接口说明如图 6-5 所示。

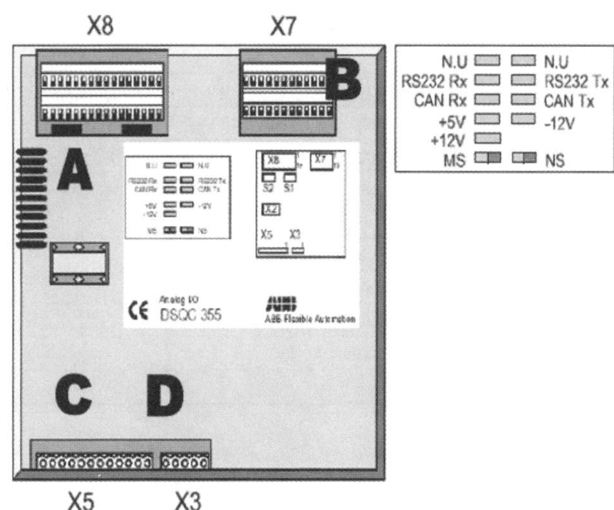

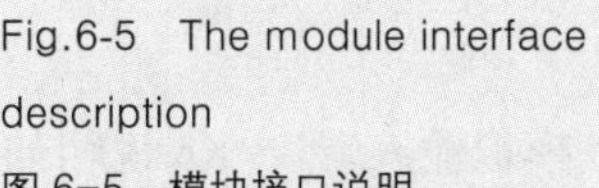

Fig.6-5 The module interface description

图 6-5 模块接口说明

其中 A 区 X8 端口为模拟输入端口，B 区 X7 端口为模拟输出端口，C 区 X5 端口为 DeviceNet 接口，D 区 X3 端口为供电电源接口。

The X8 port in area A is an analog input port, the X7 port in area B is an analog output port, the X5 port in area C is a DeviceNet interface, and the X3 port in area D is a power supply interface.

供电电源接口 X3 端口说明如表 6-9 所示。

The description of the X3 port of the power supply interface is shown in the following Tab. 6-9.

Tab. 6-9 interface X3 description

表 6-9 接口 X3 说明

X3 Terminal No.	Use Definition
1	0V
2	Unused
3	Grounded
4	Unused
5	+24V

模拟输出端口 X7 说明如表 6-10 所示。

The description of the analog output port X7 is shown in the following Tab. 6-10.

X7 Terminal No.	Use Definition	Address Allocation
1	Analog output _1 -10 V/+10 V	0-15
2	Analog output _2 -10 V/+10 V	16-31
3	Analog output _3 -10 V/+10 V	32-47
4	Analog output _4 4-20 mA	48-63
5-18	Unused	
19	Analog output _1 0V	
20	Analog output _2 0V	
21	Analog output _3 0V	
22	Analog output _4 0V	
23-24	Unused	

Tab. 6-10 interface X7 description

表 6-10 接口 X7 说明

The description of the analog input port X8 is shown in the following Tab. 6-11.

模拟输入端口 X8 说明如表 6-11 所示。

X8 Terminal No.	Use Definition	Address Allocation
1	Analog input _1 -10 V/+10 V	0-15
2	Analog input _2 -10 V/+10 V	16-31
3	Analog input _3 -10 V/+10 V	32-47
4	Analog input _4 -10 V/+10 V	48-63
5-16	Unused	
17-24	+24V	
25	Analog input _1 0 V	
26	Analog input _2 0 V	
27	Analog input _3 0 V	
28	Analog input _4 0 V	
29-30	0V	

Tab. 6-11 interface X8 description

表 6-11 接口 X8 说明

6. Operation System of Parameter Management

(1)Operation one

①Select the DeviceNet Device option, and then click Show All. As shown in the Fig. 6-6.

6. 系统参数管理操作

(1) 操作一

①选中 DeviceNet Device 选项，然后单击显示全部。如图 6-6 所示。

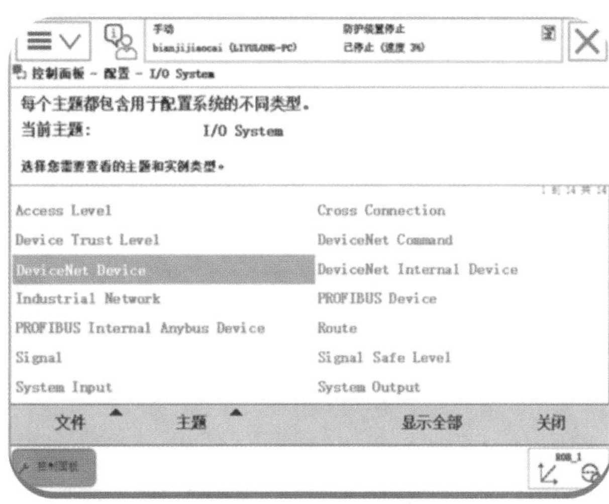

Fig.6-6 Operation Step 1

图 6-6 操作步骤 1

②选中参数实例，然后单击编辑选项。如图 6-7 所示。

②Select the parameter instance, and then click the Edit. As shown in the Fig. 6-7.

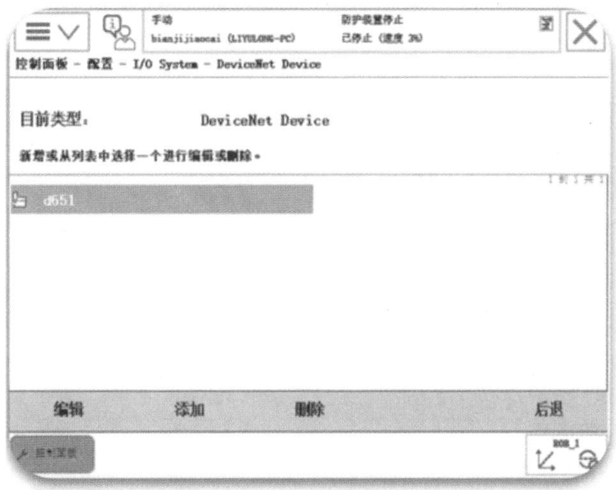

Fig.6-7 Operation Step 2

图 6-7 操作步骤 2

③对需要修改的参数名称或参数值进行双击以更改，编辑值的方法取决于值的数据类型，例如，显示下拉菜单可更改预定义值。如图 6-8 所示。

③Double-click the parameter name or parameter value that needs to be modified to change it. The method of editing the value depends on the data type of the value. For example, display the drop-down menu to change the predefined value. As shown in the Fig. 6-8.

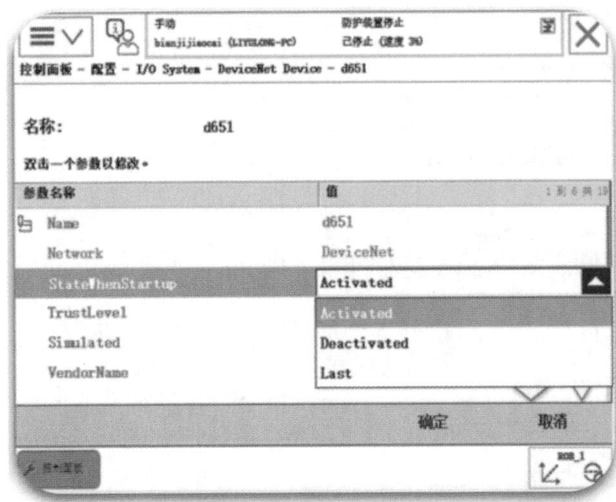

Fig.6-8 Operation Step 3

图 6-8 操作步骤 3

(2)Operation two

①Select the DeviceNet Device option and click Show All. As shown in the Fig. 6-9.

（2）操作二

①选中 DeviceNet Device 选项，单击显示全部。如图 6-9 所示。

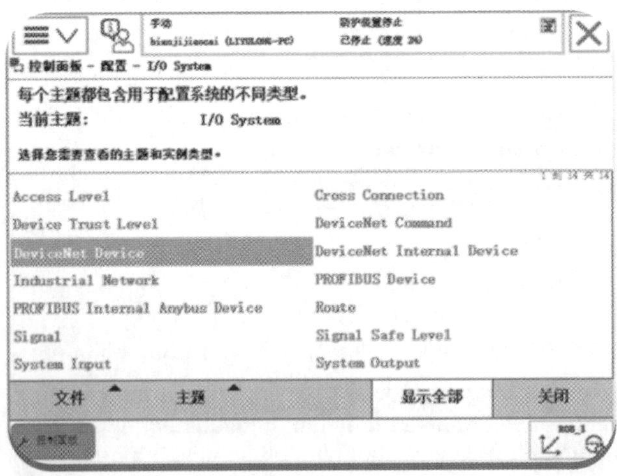

Fig.6-9 Operation Step 1

图 6-9 操作步骤 1

②Click Add to add a system parameter. As shown in the Fig. 6-10.

②单击添加，添加一个系统参数。如图 6-10 所示。

Fig.6-10 Operation Step 2
图 6-10 操作步骤 2

③单击下拉菜单，选择要添加的参数，例如，添加 DSQC 652 模块。如图 6-11 所示。

③Click the drop-down menu to select the parameter to be added, for example, add DSQC652 module. As shown in the Fig. 6-11.

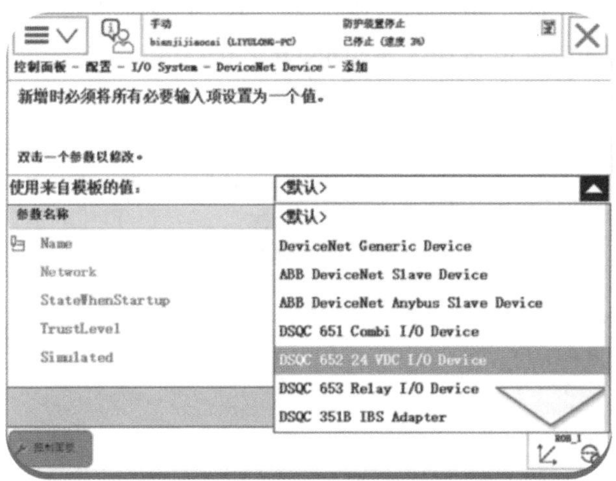

Fig.6-11 Operation Step 3
图 6-11 操作步骤 3

④若需对相关参数更改，可进行编辑；若不需修改，直接单击确定。如图 6-12 所示。

④If you need to modify the relevant parameters, you can edit it; if you don't need to modify it, just click OK. As shown in the Fig. 6-12.

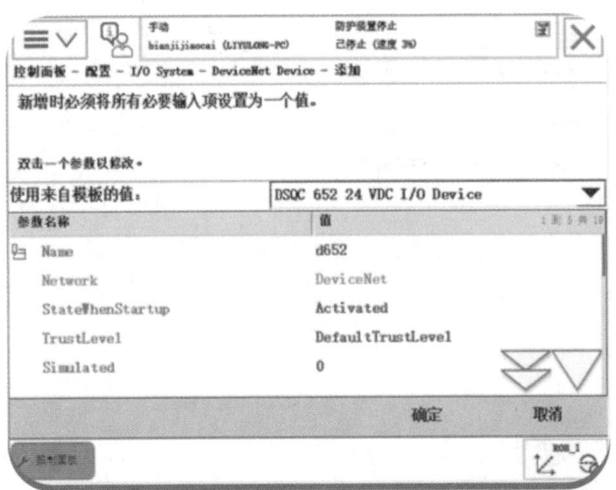

Fig.6-12 Operation Step 4

图 6-12 操作步骤 4

⑤ To take effect after the system restarts, click Yes.

⑤系统重启后生效，单击是。

When making major changes to the robot system, it is recommended to save the system parameter configuration first. As shown in the Fig. 6-13.

在对机器人系统进行较大更改时，建议先保存系统参数配置。如图 6-13 所示。

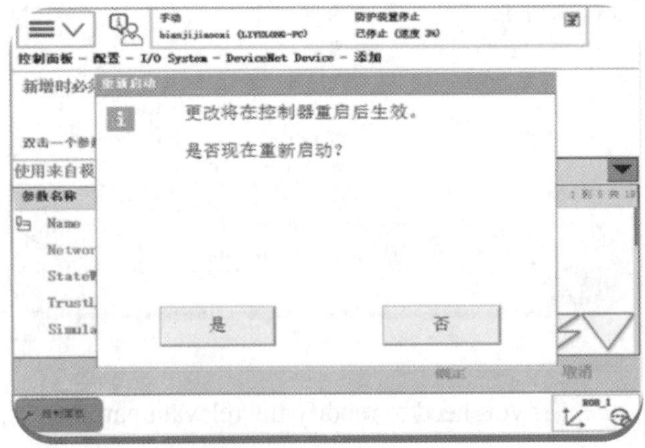

Fig.6-13 Operation Step 5

图 6-13 操作步骤 5

(3)Operation three

（3）操作三

①Select the system parameter to be saved and click the file menu. As shown in the Fig. 6-14.

①选中要保存的系统参数，单击文件菜单。如图 6-14 所示。

Fig.6-14　Operation Step 1
图 6-14　操作步骤 1

②EIO 另存为是保存已选主题的参数配置；全部另存为是保存所有主题的参数配置。例如，单击 EIO 另存为。如图 6-15 所示。

②EIO Save As is to save the parameter configuration of the selected theme; Save All As is to save the parameter configuration of all themes. For example, click EIO Save As. As shown in the Fig. 6-15.

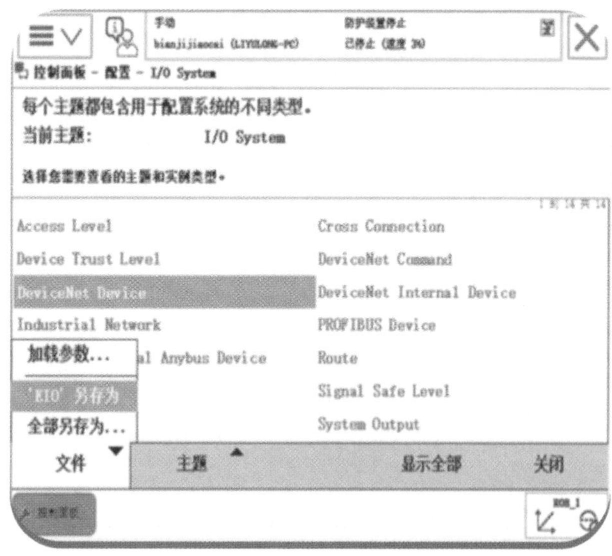

Fig.6-15　Operation Step 2
图 6-15　操作步骤 2

③选择保存参数配置的目录路径，然后单击确定。如图 6-16 所示。

③Select the directory path to save the parameter configuration, and then click OK. As shown in the Fig. 6-16.

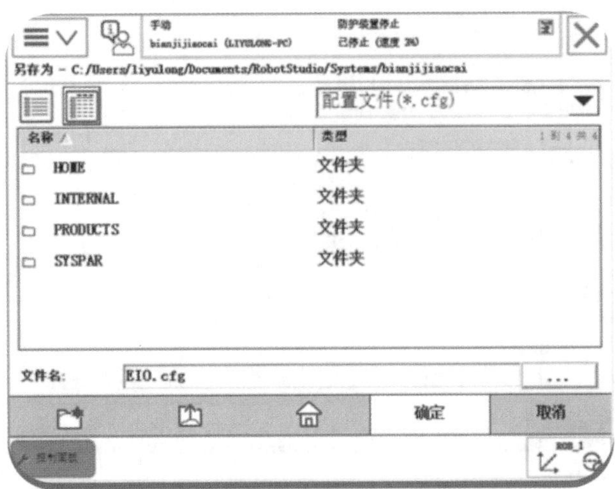

Fig.6-16 Operation Step 3

图 6-16 操作步骤 3

(4)Operation four

①In the type list, click to open the file menu. Click to load parameters. As shown in the Fig. 6-17.

（4）操作四

①在类型列表中，单击打开文件菜单，单击加载参数。如图 6-17 所示。

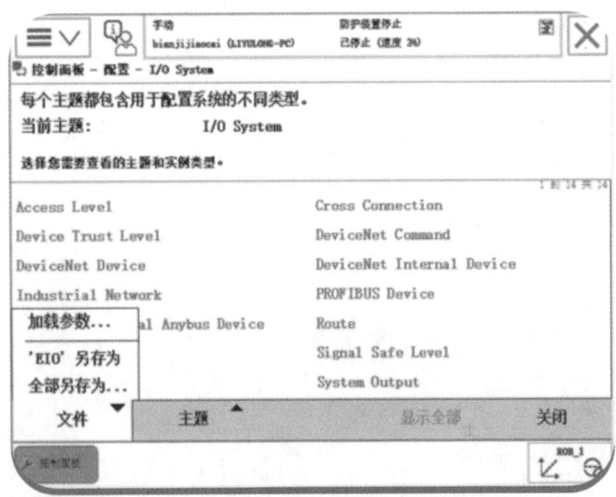

Fig.6-17 Operation Step 1

图 6-17 操作步骤 1

②Select one of the operations, and then click Load, for example, select “load parameters when there is no copy”. As shown in the Fig. 6-18.

②选择其中操作之一，然后单击加载，例如选择“没有副本时加载参数”。如图 6-18 所示。

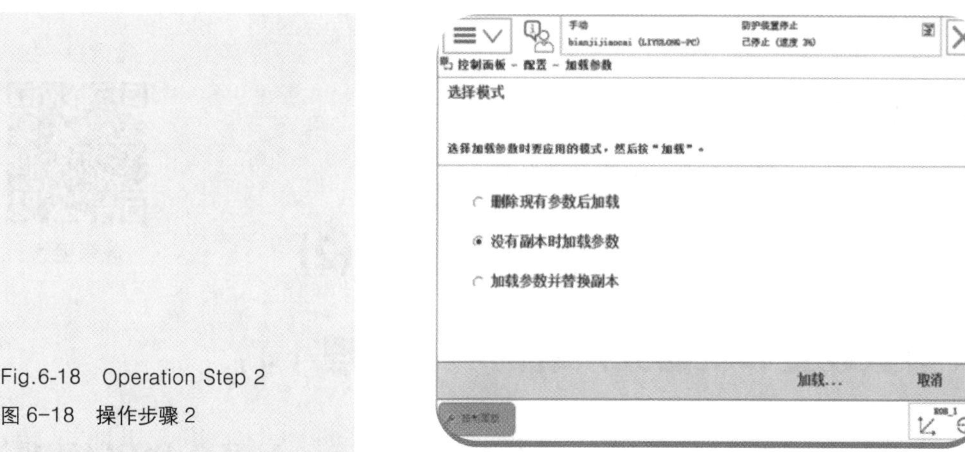

Fig.6-18 Operation Step 2

图 6-18 操作步骤 2

③选择加载参数的目录路径，然后单击确定。如图 6-19 所示。

③Select the directory path for loading parameters, and then click OK. As shown in the Fig. 6-19.

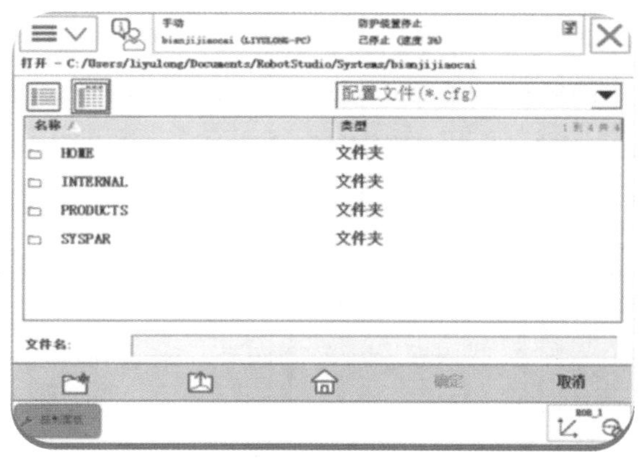

Fig.6-19 Operation Step 3

图 6-19 操作步骤 3

④单击是，系统重启后才生效。如图 6-20 所示。

④Click Yes to take effect after the system restarts. As shown in the Fig. 6-20.

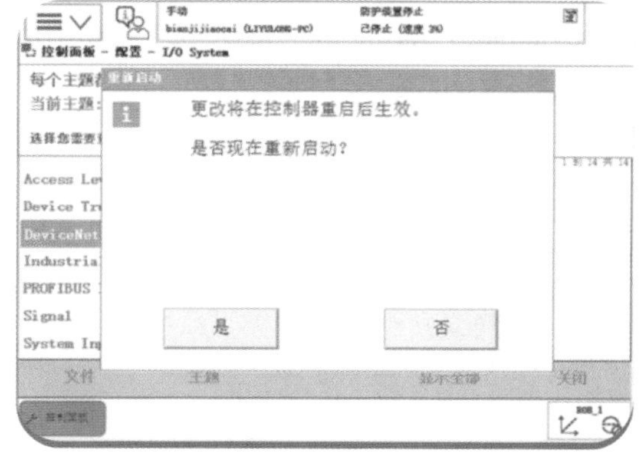

Fig.6-20 Operation Step 4

图 6-20 操作步骤 4

微信扫一扫

教学视频

Lecture 6-2

Parameter Setting and Project Management of Robot System(2)

6-2 参数设置和机器人系统的项目管理（2）

1. Define the Bus Connection of DSQC651 Board

The ABB standard I/O boards are all devices hung under the DeviceNet fieldbus, and communicate with the DeviceNet fieldbus through the X5 port.

The related parameters that define the bus connection of the DSQC651 board are described in the following Tab. 6-12.

1. 定义 DSQC651 板的总线连接

ABB 标准 I/O 板都是下挂在 DeviceNet 现场总线下的设备，通过 X5 端口与 DeviceNet 现场总线进行通信。

定义 DSQC651 板的总线连接的相关参数说明如表 6-12 所示。

Tab. 6-12 Description of relevant parameters

表 6-12 相关参数说明

Parameter Name	Set Value	Description
Name	board10	Set the name of the IO board in the system
Network	DeviceNet	IO board connected bus
Address	10	Set the address of the IO board in the bus

The steps to define the DSQC651 board in the system are as follows:

在系统中定义 DSQC651 板的操作步骤如下：

1.单击左上角主菜单按钮。

2. 选择“控制面板”。如图 6−21 所示。

1. Click the main menu button in the upper left corner.

2. Select “Control Panel” As shown in Fig. 6-21.

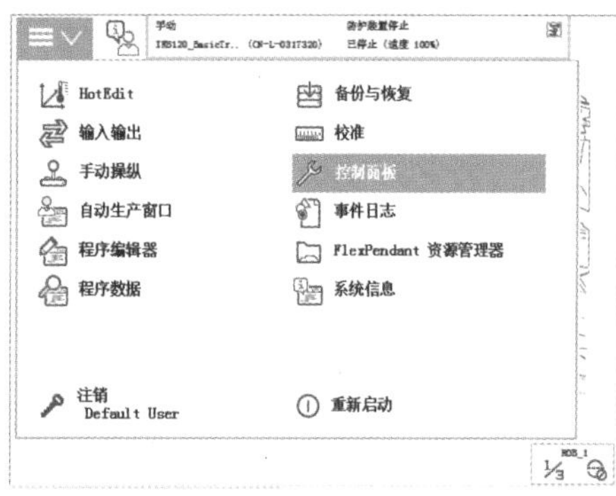

Fig.6-21 Operation Step 1 and 2

图 6−21 操作步骤 1 和 2

3. 选择“配置”。如图 6−22 所示。

3. Select “Configuration”. As shown in Fig. 6-22.

Fig.6-22 Operation Step 3

图 6−22 操作步骤 3

4. 双击“DeviceNet Device”。如图 6−23 所示。

4. Double-click “DeviceNet Device”. As shown in Fig. 6-23.

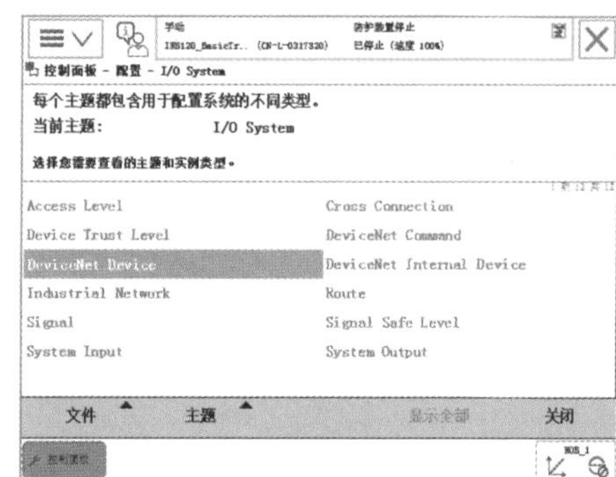

Fig.6-23 Operation Step 4

图 6−23 操作步骤 4

5. Click “Add”. As shown in Fig. 6-24.

5. 单击“添加”。如图 6–24 所示。

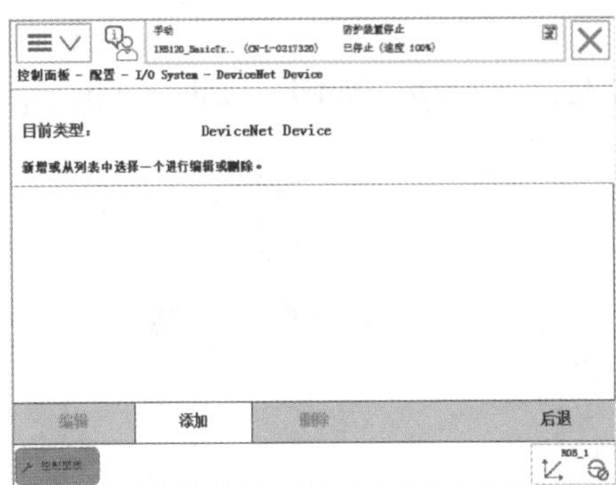

Fig.6-24 Operation Step 5

图 6–24 操作步骤 5

6. Click the drop-down arrow corresponding to “Use value from template”.

7. Select “DSQC 651 Combi I/O Device”. As shown in Fig. 6-25.

6. 单击“使用来自模板的值”对应的下拉箭头。

7. 选择“DSQC 651 Combi I/O Device”。如图 6–25 所示。

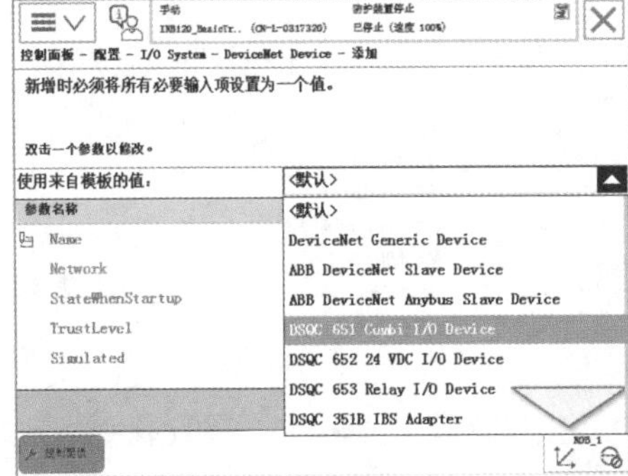

Fig. 6-25 Operation Step 6 and 7

图 6–25 操作步骤 6 和 7

8. Double-click “Name” to set the name of the DSQC651 board in the system (if it is not modified, the name is the default “d651”). As shown in Fig. 6-26.

8. 双击“Name”进行 DSQC651 板在系统中名字的设定（如果不修改，则名字是默认的“d651”）。如图 6–26 所示。

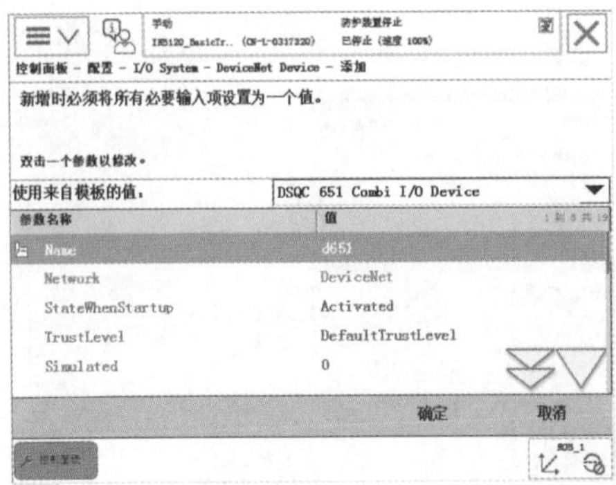

Fig.6-26 Operation Step 8

图 6–26 操作步骤 8

9. 在系统中将 DSQC651 板的名字设定为 "board10" (10 代表此模块在 DeviceNet 总线中的地址，方便识别)，然后单击 "确定"。如图 6-27 所示。

9. Set the name of the DSQC651 board in the system to "board10" (10 represents the address of this module in the DeviceNet bus for easy identification), and then click "OK". As shown in Fig. 6-27.

Fig.6-27 Operation Step 9

图 6-27 操作步骤 9

10. 单击向下翻页箭头。

11. 将 "Address" 设定为 10，然后单击" 确定"。如图 6-28 所示。

10. Click the page down arrow

11. Set "Address" to 10, and then click "OK". As shown in Fig. 6-28.

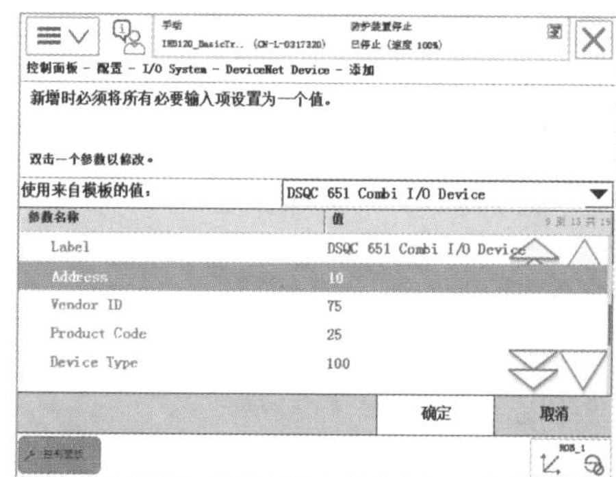

Fig.6-28 Operation Step 10 and 11

图 6-28 操作步骤 10 和 11

12. 单击 "是"，这样 DSQC651 板的定义就完成了。如图 6-29 所示。

12. Click "Yes", and the definition of the DSQC651 board is completed. As shown in Fig. 6-29.

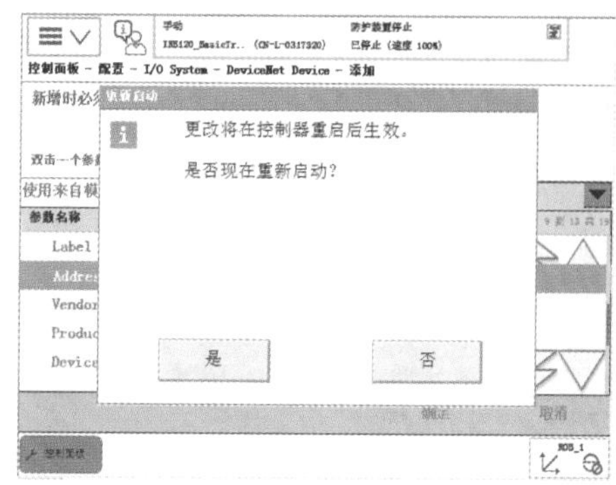

Fig.6-29 Operation Step 12

图 6-29 操作步骤 12

2. Define the Digital Input Signal di1

Define the digital input signal di1 according to the set value shown in the Tab. 6-13 below.

2.定义数字输入信号 di1

按照表 6-13 所示的设定值定义数字输入信号 di1。

Tab. 6-13 Description of signal
表 6-13 信号说明

Parameter Name	Set Value	Description
Name	di1	Set the name of the digital input signal
Type of Signal	Digital Input	Set the type of signal
Assigned to Device	board10	Set the IO module where the signal is located
Device Mapping	0	Set the address occupied by the signal

1. Click the main menu button in the upper left corner.

2. Select "Control Panel". As shown in Fig. 6-30.

1.单击左上角主菜单按钮。

2. 选择"控制面板"。如图 6-30 所示。

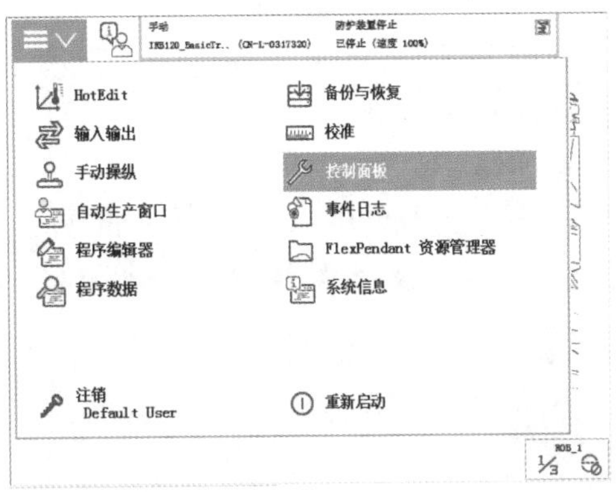

Fig.6-30 Operation Step 1 and 2
图 6-30 操作步骤 1 和 2

3.Select "Configuration". As shown in Fig. 6-31.

3. 选择"配置"。如图 6-31 所示。

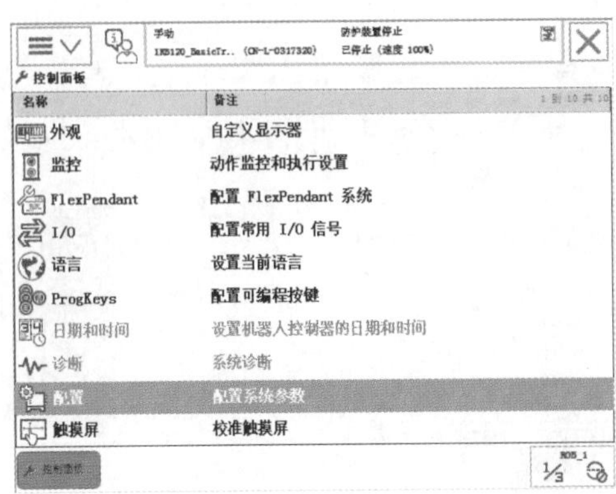

Fig.6-31 Operation Step 3
图 6-31 操作步骤 3

4. 双击 “Signal”。如图6−32所示。

4. Double-click “Signal”.As shown in Fig. 6-32.

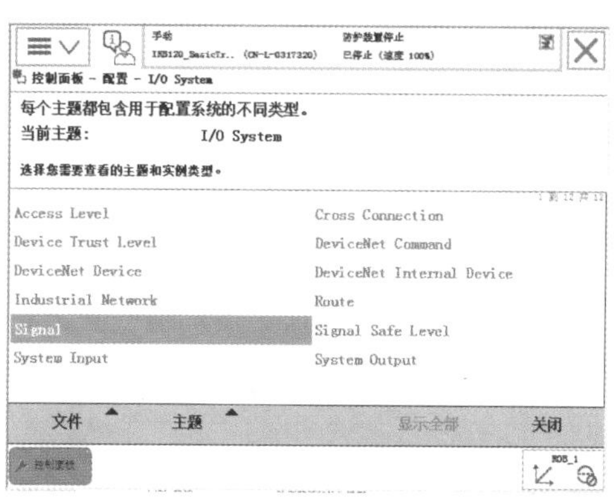

Fig.6-32 Operation Step 4

图 6−32 操作步骤 4

5. 单击“添加”。如图 6−33所示。

5. Click “Add”. As shown in Fig. 6-33.

控制面板 - 配置 - I/O System - Signal

目前类型: Signal

新增或从列表中选择一个进行编辑或删除。

ES1 ES2 SOFTESI EN1 EN2 AUTO1 AUTO2 MAN1 MANFS1 MAN2 MANFS2 USERDOOVLD MONPB AS1

编辑 添加 删除 后退

Fig.6-33 Operation Step 5

图 6−33 操作步骤 5

6. 双击 “Name”。如图 6−34所示。

6. Double-click “Name”.As shown in Fig.6-34.

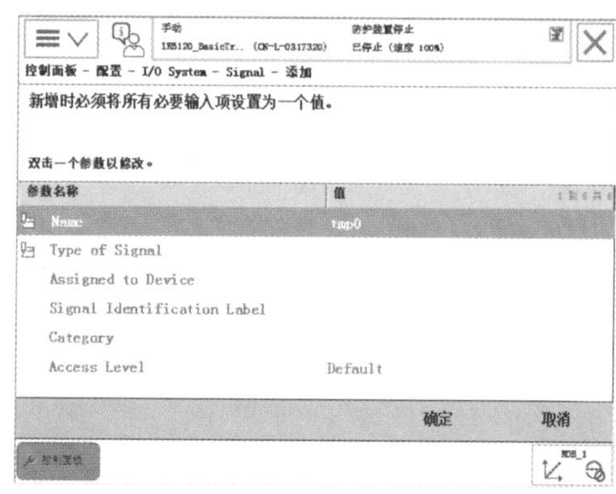

Fig.6-34 Operation Step 6

图 6−34 操作步骤 6

7. Enter “di1” and click “OK”. As shown in Fig. 6-35.

7. 输入“di1”，然后单击“确定”。如图 6–35 所示。

Fig.6-35 Operation Step 7

图 6–35 操作步骤 7

8. Double-click “Type of Signal” and select “Digital Input”. As shown in Fig. 6-36.

8. 双击“Type of Signal”，选择“Digital Input”。如图 6–36 所示。

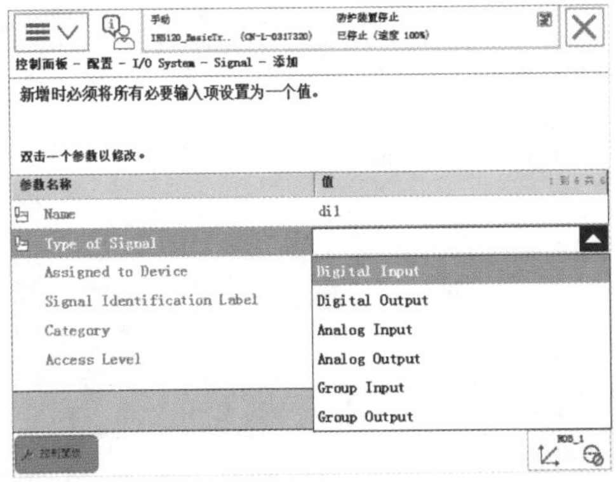

Fig.6-36 Operation Step 8

图 6–36 操作步骤 8

9. Double-click “Assigned to Device” and select “board10”. As shown in Fig. 6-37.

9. 双击“Assigned to Device”，选择“board10”。如图 6–37 所示。

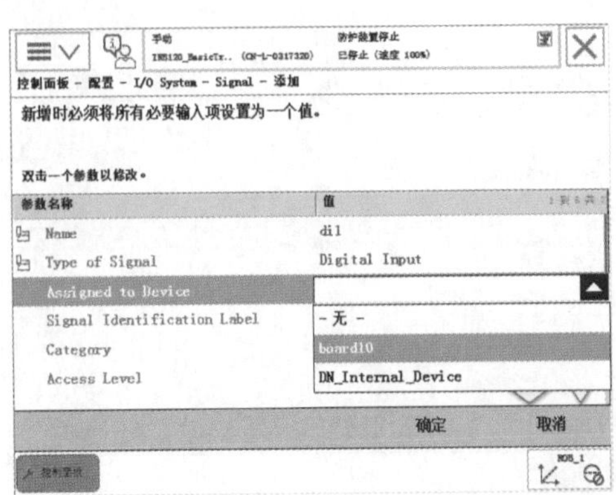

Fig.6-37 Operation Step 9

图 6–37 操作步骤 9

10.双击“Device Mapping”。如图 6−38 所示。

10. Double-click “Device Mapping”. As shown in Fig. 6-38.

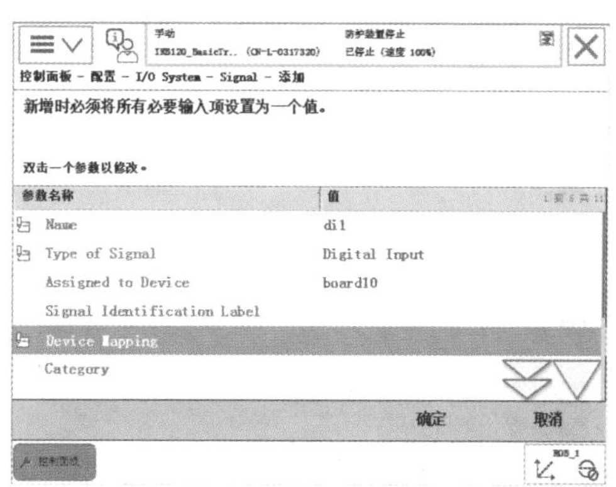

Fig.6-38 Operation Step 10

图 6−38 操作步骤 10

11. 输入 “0”，然后单击“确定”。如图 6−39 所示。

11. Enter “0” and click “OK”. As shown in Fig. 6-39.

Fig.6-39 Operation Step 11

图 6−39 操作步骤 11

12. 单击 “确定”。如图 6−40 所示。

12. Click “OK”. As shown in Fig. 6-40.

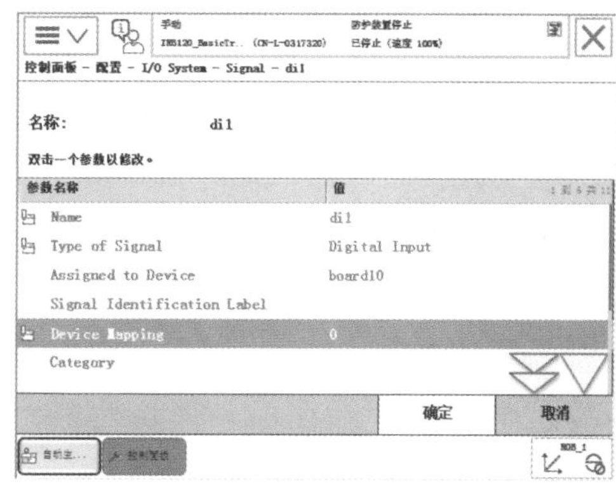

Fig.6-40 Operation Step 11

图 6−40 操作步骤 11

13. Click "Yes" to complete the setting. As shown in Fig. 6-41.

13. 单击"是"，完成设定。如图 6-41 所示。

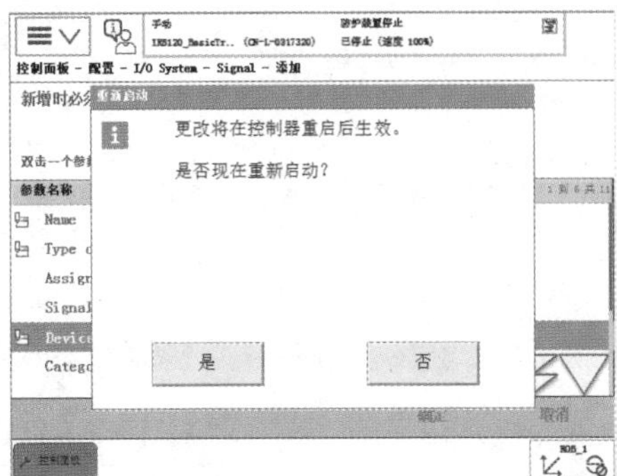

Fig.6-41 Operation Step 11

图 6-41 操作步骤 11

Exercise:

1. Define the digital output signal do;
2. Define the group input signal gi1;
3. Define the analog output signal ao.

练习：

1. 定义数字输出信号 do；
2. 定义组输入信号 gi1；
3. 定义模拟输出信号 ao。

微信扫一扫

教学视频

Lecture 7-1
Mechanical System (1)

7-1 机械系统（1）

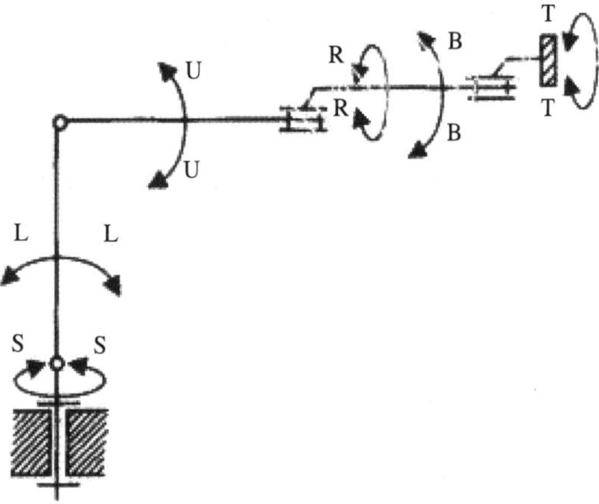

Fig.7-1 Traditional industrial robot structure

图 7-1 传统工业机器人结构

传统工业机器人的机械结构可分为主体和腕部组件两部分，如图 7-1 所示。机械臂主体连接末端执行器，而手腕控制其方向。机器人的结构由连杆和关节组成；关节运动是在两个连杆之间进行相对运动。为了便于分析，将一个连杆和与之关联的连杆看作是一个连杆对。关节运动的类型有两种：一种是转动关节产生旋转，另一种是线性或移动关节提供线性运动。至少需要六个关节才能实现对末端执行器的位置和方向的完全控制。

The mechanical structure of a conventional industrial robot can be divided into two parts, main manipulator and wrist assembly, As shown in Fig. 7-1.The manipulator is connected to the end effector while the wrist control its orientation. The structure of the robot consists of links and joints; Joint motion refers to the relative motion between. A link and its associated joint is considered as a joint-link pair for the purpose of analysis. Two types of joints are used: a revolute joint produce rotational motion, and a linear or prismatic joint provide linear motion. A minimum of six joints are required to achieve complete control of the end effector's position and orientation.

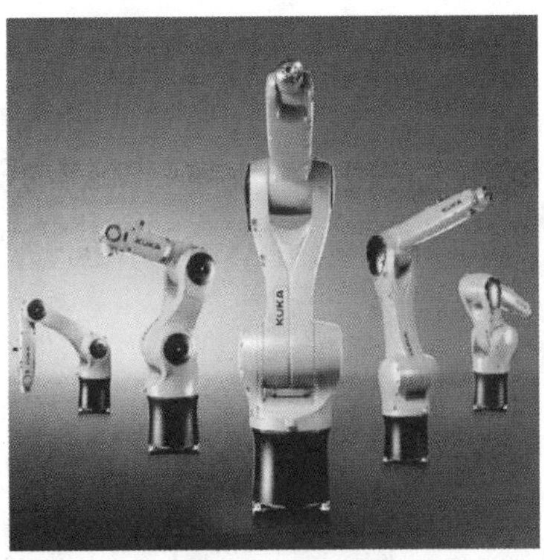

Fig.7-2 Serial robot
图 7-2 串联机器人

Most conventional industrial robots are accord with these pattern, they are known as serial robots, as shown in Fig. 7-2. serial robots often appear on assembly lines or in movies. An example is a robotic arm that helps assemble a product. It is a serial robot because the arm has joints connected in a series, and each joint has one or more motors or actuators connected to rods that enable the arm to move in what seems to be a human way. A human arm is attached by a shoulder joint first, then an elbow and finally a wrist. This is also a serial configuration, but it is a far more complex one than found in most serial robots. A person's shoulder joint allows for a wide range of movement, as does the wrist joint. To do this in robotics, several motors need to be employed with connecting rods that allow for a similar kind of movement. The "motors" in a person's arm are the muscles running throughout the arm and into the torso.

Even though many robot configurations are possible, only five configurations are commonly used within the industrial environment, As shown in Fig. 7-3:

大多数传统的工业机器人都符合这种模式，它们被称为串联机器人，如图 7-2 所示。在装配线上或在电影里经常出现串联机器人。帮助组装产品的机械臂就是一个例子。它是一个串联机器人，因为它的手臂有一系列连接在一起的关节，每个关节有一个或多个电机或驱动器连接到连杆，使手臂能够像人类一样移动。人的手臂由肩关节、肘关节和手腕连接而成。这也是一个串联结构，但它比大多数串行机器人要复杂得多。人的肩关节和腕关节都可以做大范围的运动。要在机器人技术中做到这一点，需要在连杆上安装几个电机，以实现类似的运动。手臂上的“电机”是贯穿整个手臂直至躯干的肌肉。

按照这样的设计理念，尽管有很多机器人构造方式是可行的，但在工业环境中常用的构造只有五种，如图 7-3 所示：

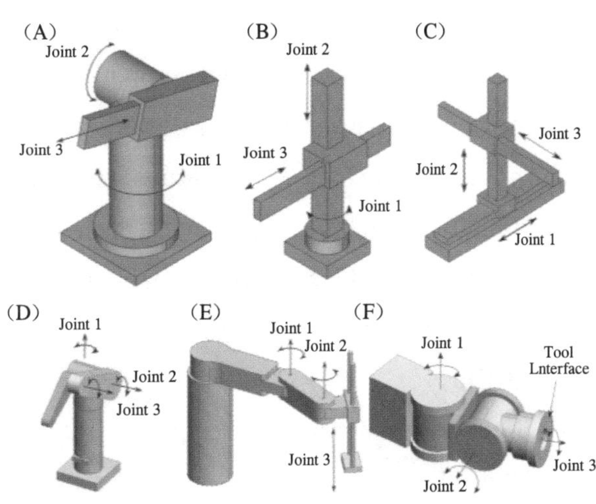

Fig.7-3 Robot construction
图 7-3 机器人构造

Polar Robot. This configuration has a linear extending arm (Joint 3) which is capable of being rotated around the horizontal (Joint 2) and vertical axes (Joint 1). This configuration is widely used in the automotive industry due to its good reach capability, Fig. 7-3(A).

极坐标型。该构型有一个可绕水平轴（关节 2）和垂直轴（关节 1）旋转的线性伸臂（关节 3）。由于其良好的伸展能力，这种结构在汽车工业中得到了广泛的应用，如图 7-3（A）所示。

Cylindrical Robot. This comprises a linear extending arm (Joint 2) which can be moved vertically up and down (Joint 3) around a rotating column (Joint 1). This is a simple configuration to control, but it has limited reach and obstacle-avoidance capabilities, Fig. 7-3(B).

柱坐标型。这包括一个可围绕旋转柱（关节 1）垂直上下移动（关节 3）的线性伸臂（关节 2）。这是一个简单的机构控制，但它只能到达有限的范围以及较低的避障能力，如图 7-3（B）所示。

Cartesian and Gantry Robot.This robot comprises three orthogonal linear joints (Joints 1-3). Gantry robots are far more rigid than the basic Cartesian configuration;they have considerable reach capabilities, and they require a minimum floor area for the robot itself, Fig. 7-3(C).

直角坐标型和龙门型。该机器人由三个正交的线性关节（关节 1-3）组成。龙门机器人远比基本的直角坐标型刚性更大；它们有相当大的伸展能力，并且只需为机器人本体提供最小的占地面积，如图 7-3（C）所示。

Revolute Robot.These robots consist of three joints (Joints 1-3) arranged in an anthropomorphic configuration. This is the most widely used configuration in general manufacturing applications, Fig. 7-3(D)

关节型。这些机器人由三个关节（关节 1–3），以拟人的方式排列。这是一般制造应用中最广泛使用的配置，如图 7-3（D）所示。

Selective-compliance-assembly robotic arm. A SCARA robot consists of two rotary axes (Joints 1 – 2) and a linear joint (Joint 3). The arm is very rigid in the vertical direction but is compliant in the horizontal direction. These attributes make it suitable for certain assembly tasks, in particular printed circuit boards, Fig. 7-3(E).

插装机械手臂型。SCARA 机器人由两个旋转轴（关节 1–2）和一个线性关节（关节 3）组成，手臂在垂直方向上具有刚性，但在水平方向上具有柔性。这些特性使它适合于某些装配任务，特别是印刷电路板，如图 7-3（E）。

A conventional robotic manipulator has three joints, this allows the tool at the end of the arm to be positioned anywhere in the robot's working envelope. To orientate the tools, three additional joints are required, these are normally mounted at the end of the arm in a wrist assembly. One design approach to a wrist is shown in Fig. 7-3(F), it must be noted that the design of the wrist can have a significant impact on the manipulator's performance, for example if the wrist has a significant mass, it will reduce the overall capabilities of the robot . The arm and the wrist give the robot the required six degrees of freedom which perm it the tool to be positioned and orientated without restrictions in three-dimension space as required by the task.

传统的机械臂有三个关节，这使得机械手末端的工具可以被放置在机器人工作范围内的任何位置。为了确定工具的方向，需要三个额外的关节，这些通常安装在手臂末端。一种手腕的设计方法如图 7-3（F）所示，必须注意手腕的设计对机械手的性能有显著的影响，例如，如果手腕负载较大，它将降低机器人的整体能力。手臂和手腕赋予机器人所需的 6 个自由度，使其能够根据任务要求在三维空间中不受限制地进行定位和定向。

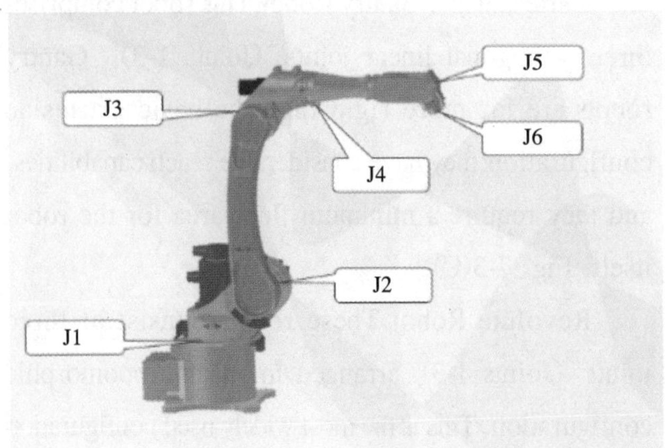

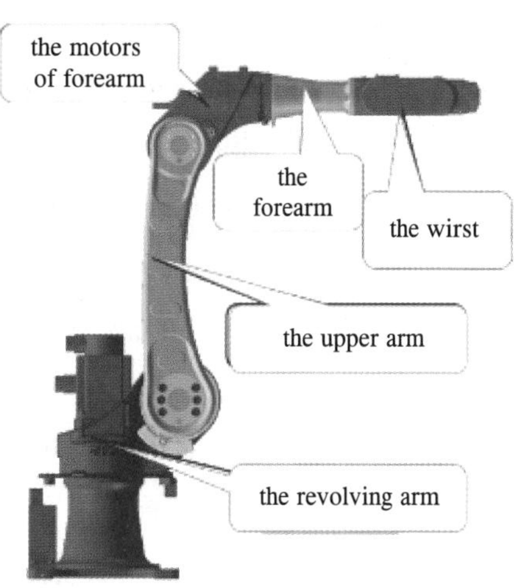

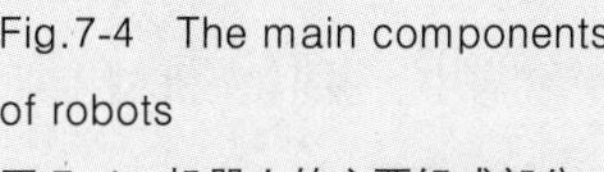

Fig.7-4 The main components of robots

图 7-4 机器人的主要组成部分

以前面提到过的工业机器人为例，该机器人有六个自由度，均为旋转关节。工业机器人本体按照组件的方式可划分为旋转臂组件、大臂组件、前臂驱动组件、前臂和腕关节组件五个主要组成部分，如图 7-4 所示。旋转臂组件与大臂组件构成工业机器人基体。

Taking the industrial robot mentioned above as an example, the robot has six degrees of freedom, all of which are rotating joints. According to the way of components, the industrial robot body can be divided into five main components: the revolving arm component, the upper arm component, the motors of forearm component, forearm and wrist component, as shown in the Fig. 7-4. The revolving arm component and the upper arm component constitute the basal body of the industrial robot.

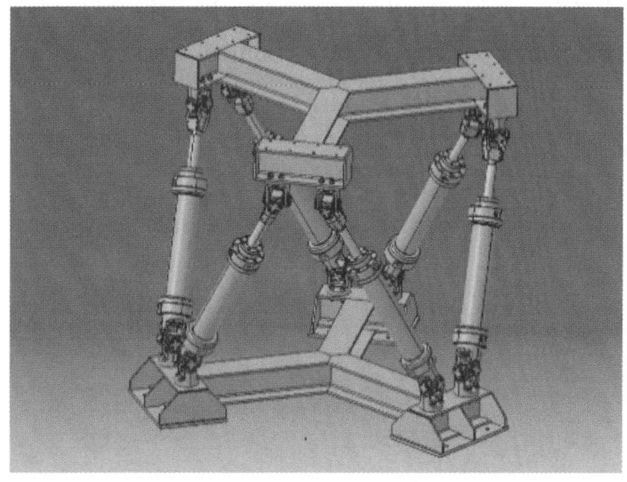

Fig.7-5 Parallel robot

图 7-5 并联机器人

Parallel robots, also called hexapods or Parallel Kinematic Machines (PKM), are closed-loop mechanisms presenting very good performances in terms of accuracy, rigidity and ability to manipulate large loads, as shown in the Fig. 7-5. parallel robots are generally faster, stronger and more accurate than their serial counterparts of similar size. They are been used in a large number of applications ranging from astronomy to flight simulators, and are becoming increasingly popular in the machinetool industry.

Closed-loop kinematic chain refers to the Rinematil chain at least one closed loop in the arrangment of connecting rods and joints.A parallel manipulator can be defined as a closed -loop mechanism composed of an end effector having n degrees of freedom (DOF) and a fixed base,linked together by at least two independent kinematic chains. Parallel mechanisms are complementary to serial mechanisms in performance. It compromised a more rigid structure with many obvious advantages over serial robot. Parallel robots own high stiffness, low inertia and relatively large payload capacity. However,Because of complex mechanism, it also has some drawbacks, such as limited workspace and nonlinear dynamics leading to control hardships.

并联机器人，有时也被称为六足机器人或并行运动学机器 (PKM)，是闭环机构，在精度、刚性和操纵大负荷的能力方面表现出非常好的性能，如图 7-5 所示。并行机器人通常比相似大小的串行机器人更快、更强、更精确。它们被广泛用于从天文学到飞行模拟器等领域当中，并且在机械工业中越来越受欢迎。

闭环运动链是指连杆和关节的布置至少存在一个闭环的运动链。一个并联机构可以定义为由一个具有 n 个自由度的末端执行器和一个固定基座组成的闭环机构，由至少两个独立的运动链连接在一起。并联机构在性能上是串行机构的补充。它牺牲了较强的刚性结构，但有许多明显优于串行机构人的特点。并联机器人具有高刚度、低惯性和较大的载荷能力。但由于机理复杂，也存在工作空间有限、动力学非线性等问题，导致控制难度大。

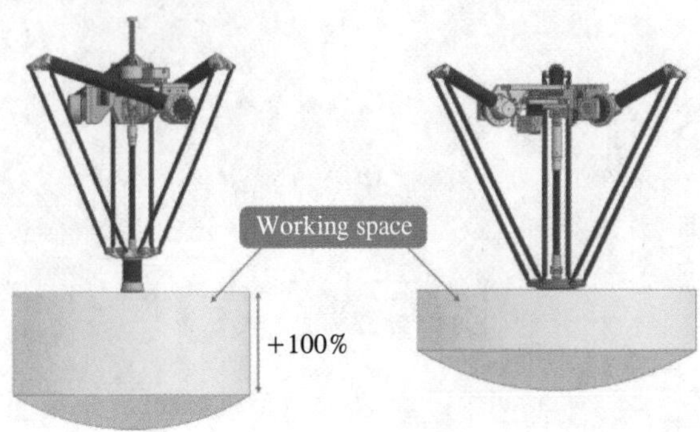

Fig. 7-6 Workspace of parallel robot

图 7-6 并联机器人的工作空间

近年来，推出了许多类型的并行机器人。虽然大多数现有的架构是基于设计者的直觉，但是并行机制的综合可以被系统地处理。Delta 机器人是应用最广泛的并联机器人之一。三角结构的关节和连杆的几何排列提供了平台上的三个平动自由度。Delta 并联机器人设计背后的基本思想是使用平行四边形。平行四边形允许输出连杆相对于输入连杆保持固定方向。三个平行四边形的使用完全限制了移动平台的方向，移动平台只剩下三个纯平动的自由度。三个平行四边形的输入杆通过转动关节安装在转动杆上。转动杠杆的转动关节以两种不同的方式驱动: 旋转（直流或交流伺服）电机或线性驱动器。最后，第四根输出连杆用于从基座向安装在移动平台上的末端执行器传输旋转运动。

使用基座安装的执行机构和低质量链接可以使移动平台实现高达 50 G 的实验环境中的加速度和 12 G 的工业应用中的加速度。这使得 Delta 机器人（如图 7-7）成为分拣和摆放轻物体（从 10 g 到 1 kg）操作的完美选择。理想情况下，它的工作空间是三个右圆环的交点。市场上的 Delta 机器人通

Numerous types of parallel robots have been proposed in the recent years. Although most existing architectures are based on the intuition of their designer, the synthesis of parallel mechanisms can be dealt with systematically. Delta robot is one of the most widely used parallel robots. The geometric arrangement of the joints and links of the Delta structure provides three translational degrees of freedom at the platform. The basic idea behind the Delta parallel robot design is the use of parallelograms. A parallelogram allows an output link to remain at a fixed orientation with respect to an input link. The use of three such parallelograms restrain completely the orientation of the mobile platform which remains only with three purely translational degrees of freedom. The input links of the three parallelograms are mounted on rotating levers via revolute joints. The revolute joints of the rotating levers are actuated in two different ways: with rotational (DC or AC servo) motors or with linear actuators. Finally, a fourth leg is used to transmit rotary motion from the base to an end-effector mounted on the mobile platform.

The use of base-mounted actuators and low-mass links allows the mobile platform to achieve accelerations of up to 50 G in experimental environments and 12 G in industrial applications. This makes the Delta robot (shown in Fig.7-7) a perfect candidate for pick and place operations of light objects (from 10 g to 1 kg). Ideally, its workspace is the intersection of three right circular tori. The Delta robots available on the market operate typically in a cylindrical workspace which is 1 m in diameter and 0.2 m high.

常在直径 1 米、高 0.2 米的圆柱形工作空间中工作。

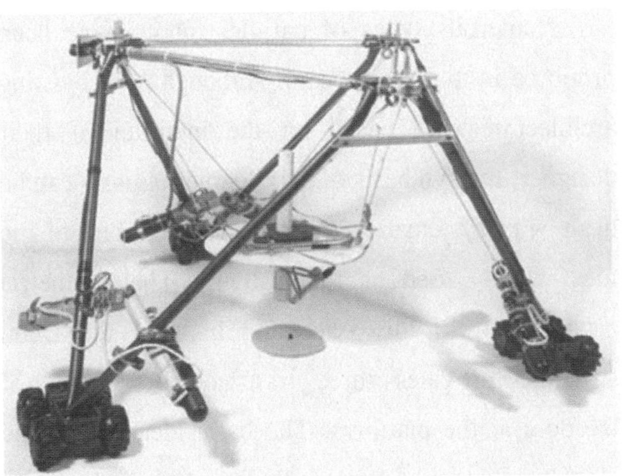

Fig.7-7 Delta robot
图 7-7 三角机器人

Parallel mechanisms naturally lead themselves to another paradigm that exploits the structural properties of cables shown in Fig.7-8. Cables are mechanical components that can with stand large tensile loads relative to their weight. When applied to parallel mechanisms, a plurality of cables can be attached to a moving platform and driven by actuated reels mounted on the base frame.

并行机制自然会导致另一个范例的出现，利用绳索的结构特性，如图 7-8 所示。绳索是一种能够承受相对于自身重量的较大拉力载荷的机械部件。当应用于并联机构时，可将多个电缆连接到移动平台上，并由安装在基础框架上的驱动卷轴驱动。

Because of the parallel architecture, it is possible to ensure that cables can be maintained under tension with in a given work space. The advantages of these mechanisms are numerous . When wound on a spool, cables allow motion ranges that are much wider than that produced by conventional articulated systems. Also, because they can only resist to tensile forces, cables are, as mentioned above, much thinner and lighter than most conventional mechanical components, for similar pay loads. Thus, they have very low inertia and are particularly suitable for systems in which large ranges of motion are required. Indeed, long-reach a rticulated robots are heavy and involve the motion of

由于采用并联结构，绳索能够给定的工作空间中保持在张力下。这些机构的优点很多。当缠绕在线轴上时，绳索允许的运动范围比传统的铰接系统产生的运动范围大得多。此外，由于它们只需要抵抗拉力，因此，如前所述，对于类似的有效载荷，钢索比大多数传统机械部件要轻巧得多，也轻得多。因此，它们具有非常低的惯性，特别适用于需要大范围运动的系统。事实上，长距离关节式

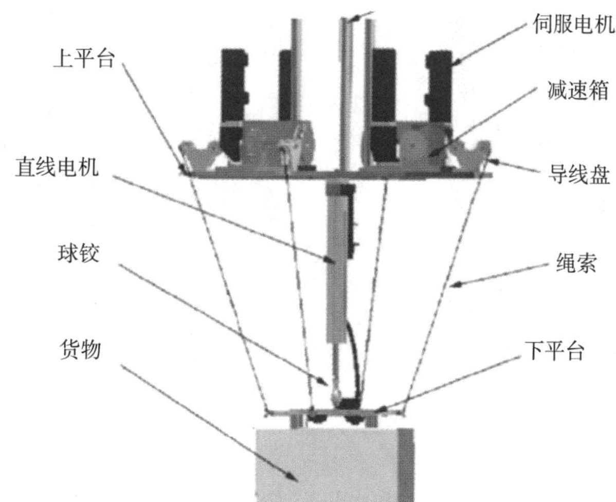

Fig.7-8 The structural properties of wires

图 7-8 绳索结构

significant mass, whereas cable-driven robots remain light even when the range of motion is very large.

Because wires can only resist tensile loads, wire-driven parallel mechanisms are generally designed based on either one of the two following approaches:

(1) the weight (gravity) of the moving platform is used to ensure that all cables are under tension or (2) the number of cables is larger than the number of degrees of freedom of the platform, thereby allowing a pre-stressing of the cables that can be used to ensure that all cables remain under tension. The wire-driven parallel mechanisms that are designed based on the first approach are referred to as cable-suspended parallel mechanisms, while the mechanisms that are based on the second approach are referred to as fully constrained wire-driven parallel mechanisms.

机器人很重，涉及大量的运动，而缆绳驱动的机器人即使在运动范围非常大的情况下仍然很轻。

由于绳索只能承受拉伸载荷，绳索驱动并联机构的设计一般基于以下两种方法之一：

（1）移动平台的重量（重力）用于确保所有电缆处于张力下，或（2）电缆的数量大于平台的自由度，从而允许对电缆施加预应力，以确保所有电缆处于张力下。基于第一种方法设计的线驱动并联机构称为悬索并联机构，而基于第二种方法设计的机构称为完全约束线驱动并联机构。

Fig.7-9 Biomimetics
图 7-9 仿生学

Biomimetics (shown in Fig.7-9) is a broad field that covers all ranges of robotics including robot structure and mechanics, actuation, perception, and autonomy. Robots that mimic structure and movement principles found in nature to perform desired tasks in unstructured environments.

仿生学（如图 7-9）是一个广泛的领域，涵盖了机器人结构和力学、驱动、感知和自主性的所有领域。模仿自然界结构和运动原理的机器人，以便在非结构化环境中执行所需任务。

Nature frequently inspires engineers to adopt solutions from biology for application to human challenges. Robots are built to perform certain tasks, and many tasks include moving. The various modes of locomotion in nature have inspired robots to be made to mimic these locomotion with a goal to overcome various obstacles in the environment, and move around with the extreme agility similar to that found in nature. Mankind has engineered various modes of transportation on ground, air, and water. On ground, wheeled vehicles are the most popular choice. In air, fixed wing air planes an d helicopters with rotating blades dominate, and in water, ships, and submarines propelled by similar rotating elements are most common. In contrast, nature has different solutions to locomotion involving moving legs, undulating fins, and flapping wings. Many of these

大自然经常激励工程师们采用生物学的解决方案来应对人类的挑战。机器人是用来执行某些任务的，而任何任务都包括两地间的移动。自然界中各种各样的运动模式激发了机器人模仿这些运动的灵感，目的是克服环境中的各种障碍，并以与自然界相似的极端敏捷性四处移动。人类在地面、空中和水上设计了各种运输方式。轮式车辆是地面上最常见的；固定翼飞机和带有旋转叶片的直升机在空中占主导地位；由类似旋转部件推动的船舶和潜艇在水中最为常见。相比之下，大自然对运动有不同的解决方案，包括移动的腿、起伏的鳍和拍打的翅膀。许多这种生物运动机制与工程系统有相似之处。大自然不使用轮子，而是使用不同大小、数量和阻抗的躯干；人类和鸟类是两足动物，许多哺乳动物和爬行动物用四条腿活动，昆虫有六条腿，其他节肢动物有八条腿或更多。这些躯干在不同的表面上协调地以稳定的方式移动。蛇和蠕虫通过身体产生曲折来移动。鸟类、蝙蝠和飞虫通过拍打翅膀飞行，区别于喷气式飞机和旋转螺旋桨。鱼用扭动身体的运动方式游动，其灵敏性远超传统船只和潜艇。每一种生物运动模式都是基于自然界的基本促动器：肌肉。肌肉带动躯干产生线性运动。因此，躯干运动与肌肉的形态密切相关。

人造机器移动的主要目的是在尽可能短的时间内或以最小的能量消耗实现长距离运送大量载荷。公路、机场、造船厂和其他基础设施支持这种运输，使汽车、火车、飞机和船舶的结构设计能够集中于运输任务。另一方面，自然界的运动主要是为了生存。动物和昆虫已经进化到能在各种环境中生存下来。每一种动物都利用

biological locomotion mechanism shave analogies to engineered systems. Instead of using wheels, nature uses legs of different sizes, numbers,and impedances; humans and birds are bipedal, many mammals and reptiles move on four legs, and insects have six legs, and other arthropods have eight or more legs. These legs are coordinated to move on different surfaces in a stable manner. Snakes and worms move around without legs by creating waves with their body. Instead of fixed wings with jets and rotary propellers, birds , bats, and flying insects flap their wings. Fish create undulatory motions with their bodies to swim with agility far beyond conventional boats and submarines. Each of these biological locomotion modes are based on the nature's fundamental actuator: muscle. Muscles create linear motion that is coupled to the structures that generate locomotion. Therefore, these structures are closely related to the morphology of the muscles that move the structures.

The main purpose of locomotion in manmade machines is to deliver large payloads across long distances in the shortest possible time or with the minimal energy expenditure. There are roads, airports, shipyards, and other foundations that support this transportation, enabling the structural design of cars, trains, planes, and ships to focus on transportation tasks. On the other hand, locomotion in nature is primarily for survival. Animals and insects have evolved to survive in various environments. Each species uses their mode of locomotion to hunt for food, find mates, and escape danger all in unstructured natural environments. Therefore, the requirements for biological locomotion are much more complicated than human transportation systems.

它们的移动模式来寻找食物、寻找配偶和逃离危险，所有这些都是在非结构化的自然环境中进行的。因此，生物运动比人类交通系统要复杂得多。

Fig.7-10 Biomimetic robots
图 7-10 仿生机器人

Biomimetic robots (shown in Fig.7-10) try to mimic the structural characteristics and the principles of movement to be able to move around places where conventional machines or robots are unable to perform as needed. Animals can crawl on a rugged terrain at a high speed, can climb walls without a tether, can fly in cluttered environments and hover and perch as needed .

仿生机器人(如图 7-10 所示)模仿结构特征和运动原理，以便在传统机器或机器人无法按需要执行的地方移动。动物可以在崎岖不平的地形上高速爬行，可以不用系绳爬墙，可以在杂乱的环境中飞行，可以在需要的时候盘旋和栖息。

Depending on the size of the species the optimal mode of locomotion and the underlying structure is different. For example, jumping is found frequently in small insects to escape danger, since their small size makes it hard for them to escape quickly using other forms of locomotion. Jumping by a large animal is different from jumping by small insects like a flea. Larger insects or animals tend to run or crawl to escape the danger. Large birds flap their wings at a much lower frequency, and use gliding mode whereas

根据物种的大小不同，最佳的移动方式和内部的结构也是不同的。例如，小昆虫经常通过跳跃来逃避危险，因为它们的体形太小，很难用其他的移动方式快速逃脱。大型动物跳跃和跳蚤之类的小昆虫跳跃是不同的。较大的昆虫或动物倾向于用它们的腿跑或爬行来逃避危险。大型鸟类扇动翅膀的频率要低得多，它们使用滑翔模式，而蜜蜂和苍蝇在飞行过程中不断地以高频率扇动翅膀。鸟类的翅膀有骨骼、肌肉和羽毛，而昆虫的翅膀没有肌肉，重量很轻。因此，在开发仿生机器人时，需要考虑目标种群的大小。

bees and flies beat their wings continuously at a high frequency during flight. The structure is very different, where the wings of a bird have bones, muscle, and feathers versus the muscle-less and lightweight wings of insects. Therefore, when developing biomimetic robots, the size of the target species should be taken into account.

Fig.7-11 The wall climbing robot Stickybot

图 7-11 壁虎爬墙机器人

动物在大自然中不断进化以适应极端环境，所以有很多极端运动的例子，这在传统的工程设计中是不可能实现的。例如，壁虎爬墙机器人 Stickybot（如图 7-11）利用壁虎的定向附着原理爬上光滑的垂直墙壁。然而，仅靠定向附着是不够的。这种设计必须使压垫上的压力分布均匀，以确保压垫与壁面的良好接触。这些小细节对仿生机器人的性能很重

Animals constantly evolve in nature to adapt to extreme environments, there are many examples of extreme locomotion that are typically not possible with conventional engineering designs. For example, the wall climbing robot Stickybot (shown in Fig.7-11) uses the directional adhesion principle of a gecko to climb up smooth vertical walls. However, directional adhesion alone is not enough. The design has to enable even pressure distribution on the pad to make sure the pads are well in contact with the wall. These small details can be important to performance of biomimetic robots and should be considered carefully. Since robots are commonly built to perform tasks too tedious or dangerous for humans, it is natural to adopt designs found in nature for functionality where nature has already found a solution.

要，应该仔细考虑。由于机器人通常被制造来执行对人类来说过于单调或危险的任务，自然会采用自然界已经找到解决方案的设计来实现功能。

Lecture 7-2

Mechanical System (2)

7-2 机械系统（2）

教学视频

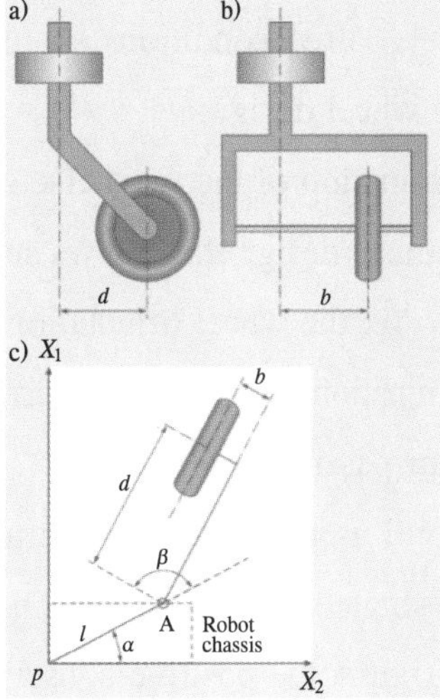

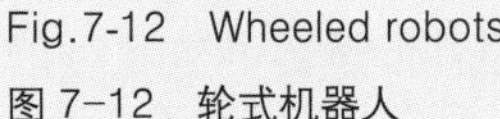

Fig.7-12 Wheeled robots

图 7-12 轮式机器人

轮式机器人（如图 7-12 所示）由于其结构简单、节能、速度快、制造成本低等优点，被广泛应用于移动领域。

Wheeled robots (shown in Fig.7-12) have been widely used to achieve mobility because there are many advantages including the simple structure, energy efficiency, fast speed, low fabrication cost, and so forth.

In order to achieve robot locomotion, wheeled mobile robots are widely used in many applications. In general, wheeled robots consume less energy and move faster than other locomotion mechanisms (e.g., legged robots or tracked vehicles). From the viewpoint of control, less control effort is required, owing to their simple mechanisms and reduced stability problems. Although it is difficult to overcome rough terrain or uneven ground conditions, wheeled mobile robots are suitable for a large class of target environments in practical applications. In the single, wheel design scheme, there are two candidates: a standard wheel or a special wheel. standard wheel can be understood as conventional tire. Special wheels possess unique mechanical structures including rollers or spheres. Fig. 7-12 shows the general design of a standard wheel. Three conditions should be defined for a standard wheel design:

1. Determination of the two offsets d and b;

2. Mechanical design that allows steering motion or not (i.e., to fix the wheel orientation or not);

3. Determination of steering and driving actuation (i. e., active or passive drive).

Condition 1 is the kinematic parameter design problem for a single standard wheel. The parameter d can be either 0 or some positive constant. Parameter b is the lateral offset of the wheel and is usually set to zero. In a special design, a nonzero b may be s elected to obtain pure rolling contact between the wheel and ground without causing rotational slip at the contact point. However, this is rarely used and we mainly consider the case of zero lateral offset b.

为了实现机器人的运动，轮式移动机器人在许多领域得到了广泛的应用。一般来说，轮式机器人消耗的能量更少，比其他运动机构（如腿式机器人或履带式车辆)移动得更快。从控制的角度看,其机构简单,稳定性问题少,控制工作量小。虽然很难克服崎岖的地形或不平整的地面条件，但轮式移动机器人在实际应用中适用于大多数的环境。在单轮设计时，有两种选择:标准轮或特殊轮。标准轮胎可以理解为一个常规轮胎。特殊车轮具有独特的机械结构，包括滚轮或球体。图中显示了标准轮的一般设计。对于标准的车轮设计，应该定义三个条件：

1.确定两个偏移量 d 和 b;

2. 机械结构是否允许转向运动（车轮定向);

3. 转向和驱动的决定(即主动或被动驱动)。

条件 1 是单标准轮的运动学参数设计问题。参数 d 可以是 0，也可以是某个正常数。参数 b 是车轮的横向偏移量，通常设置为零。在特殊的设计中，可以选择非零 b，以获得车轮与地面之间的纯滚动接触，而不会在接触点造成转动滑移。但是这种情况很少使用，主要考虑的是零侧向偏移量 b 的情况。

条件 2 是车轮方向是否可以改变的设计问题。如果转向轴是固定的，车轮提供了一个速度约束的驱动方向。条件 3 是由执行器驱动转向或驱动运动，还是被动驱动转向或运动的设计问题。

在允许转向运动的情况下，偏置 d 在运动学建模中起重要作用。对于常规的脚轮(偏心的定向轮)，有一个在零偏移 d。图中的点 A 表示连接轮毂模块到机器人底盘的关节位置。通过车轮模块的转向和驱动运动，得到了 A 点处两个正交的线速度分量。这意味着被动的脚轮不会对机器人的运动提供额外的速度约束。如果脚轮装有两个独立进行转向和驱动运动的制动器，通过求解逆运动学问题可以得到 A 点的任意期望速度，因此可以实现完整的全向运动。

如果偏移量 d 被设置为零，点 A 的允许速度方向被限制在车轮方向上。在这种情况下，转向运动不应该是被动的，因为车轮的方向不能被动地改变。然而，驱动速度可以被动地由其他轮的驱动来决定。

由于非完整速度约束，车

Condition 2 is a design problem for whether the wheel orientation can be changed or not. If the steering axis is fixed , the wheel provides a velocity constraint on the driving direction . Condition 3 is the design problem of whether to actuate steering or driving motion by actuators or to drive steering or motion passively.

If steering motion is allowed , the offset d plays a significant role in the kinematic modeling. For a conventional caster wheel (an off centered orientable wheel), there is an on zero offset d. Point A. indicates the location of the joint connecting the wheel module to the robot chassis. Two orthogonal linear velocity components at point A are obtained by a casterwheel, which results from the steering and driving motions of the wheel module. This implies that a passive caster wheel does not provide an additional velocity constraint on the robot's motion. If a caster wheel is equipped with two actuators that drive steering and driving motions independently, holonomic omnidirectional movement can be achieved because any desired velocity at point A can be generated by solving the inverse kinematics problem .

If the offset d is set to zero, the allowable velocity direction at point A is limited to the wheel orientation. In such a case, the steering motion should not be passive because the wheel orientation cannot be changed passively. However, the driving velocity can be determined passively by the actuation of other wheels.

Wheel orientation should be actively steered to the desired velocity direction due to the nonholonomic velocity constraint. This implies that the orientation of wheels should be aligned before movement.

In summary, four types of standard wheels are commonly used.First is a passively driven wheel with a fixed steering axis. Second is a passive caster wheel with offset d . Third is an active caster wheel with offset d, where the steering and driving motions are controlled by actuators. The fourth is an active orientable wheel with zero offset d , where steering and driving motions are driven by actuators. The structures of each wheel type are shown in Fig .

轮的方向应该主动转向到期望的速度方向。这意味着在移动之前，车轮的方向应该相同。

总之，常用的标准车轮有四种。首先是一个被动驱动的车轮与固定的转向轴。二是带偏置 d 的被动脚轮。第三是主动脚轮与偏置 d，其中转向和驱动运动是由执行机构控制。第四是主动定向车轮与零偏移 d,其中转向和驱动运动是由执行机构驱动。

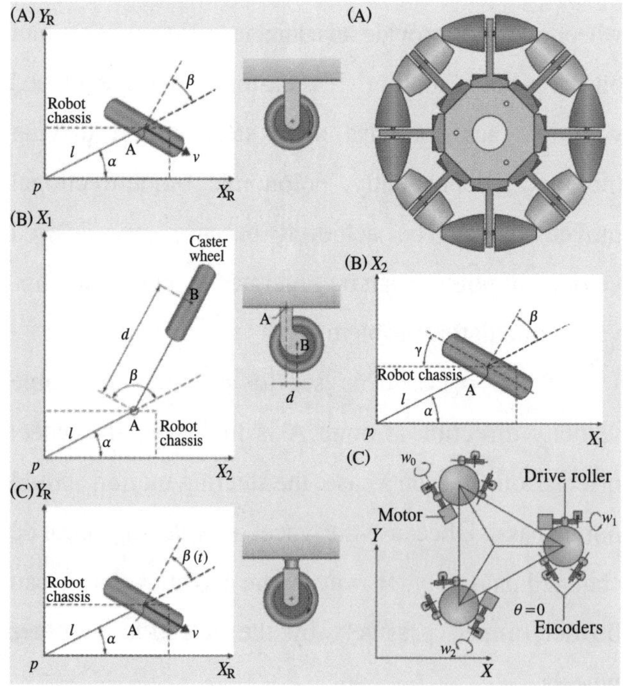

Fig.7-13 Swedish wheel

图 7-13 瑞典车轮

虽然标准轮具有结构简单、可靠性好等优点，但非完整速度约束（无侧滑条件）限制了机器人的运动。另一方面，可以使用特殊的轮子来实现移动机器人（全移动机器人）的全向运动，即保证平面运动的三个自由度。我们考虑了两种典型的特殊车轮设计：瑞典车轮和球形车轮。

图 7-13 中即是瑞典车轮。小的被动自由滚轮位于车轮的外缘。采用自由滚轮以消除非完整速度约束。被动滚轮可以自由地绕转轴旋转，从而产生车轮的横向运动。因此，在控制驾驶速度时，横向速度是由其他车轮的驱动方向决定的。

如图 7-13（c）所示，球形轮的旋转受到与球形接触的滚轮的约束。托辊可分为驱动托辊和支撑托辊。球体由驱动滚轮驱动,尽管滚动接触提供了非完整的约束,球体模块的最终运动变成了完整约束。这意味着机器人可以在任何时间以任何期望的线速度/角速度移动。

利用球形轮可以研制出一种完整的全向移动机器人，实现球体与地面的光滑连续接触。然而，球支承机构的设计难度较大，而且由于点接触的原因，其载荷必须很低。另一

Although standard wheels are advantageous because of their simple structure and good reliability, the nonholonomic velocity constraint (no side-slip condition) limits robot motion. On the other hand, special wheels can be employed in order to obtain omnidirectional motion of a mobile robot (omnimobile robot), to ensure three degrees of freedom for plane motion. We consider two typical designs of special wheels: the Swedish wheel and the spherical wheel.

Fig. 7-13 showed Swedish wheel. Small passive free rollers are located along the outer rim of the wheel. Free rollers are employed in order to eliminate the nonholonomic velocity constraint. Passive rollers are free to rotate around the axis of rotation, which results in lateral motion of the wheel. As a result, a driving velocity should be controlled, while the lateral velocity is passively determined by the actuation of the other wheels.

A spherical wheel is shown in Fig.7-13(c). The rotation of the sphere is constrained by rollers that make rolling contact with the sphere. The rollers can be divided into driving and supporting rollers. The sphere is driven by the actuation of the driving rollers, whereas the rolling contacts provide nonholonomic constraints, and the resultant motion of the sphere module becomes holonomic. This implies that the robot can be moved with any desired linear/angular velocities at any time.

By using the spherical wheel, a holonomic omnidirectional mobile robot can be developed and the robot achieves smooth and continuous contact between the sphere and the ground. However, the design of the sphere-supporting mechanism is difficult and the pay-load must be quite low due to the point contact. Another drawback is that the surface of the sphere can be polluted when traveling over dirty ground and it is difficult to overcome irregular ground conditions. These draw-backs limit the practical application of the spherical wheel.

个缺点是，球面在肮脏的地面上行驶时可能会受到污染，而且很难克服不规则的地面条件。这些缺点限制了球轮的应用。

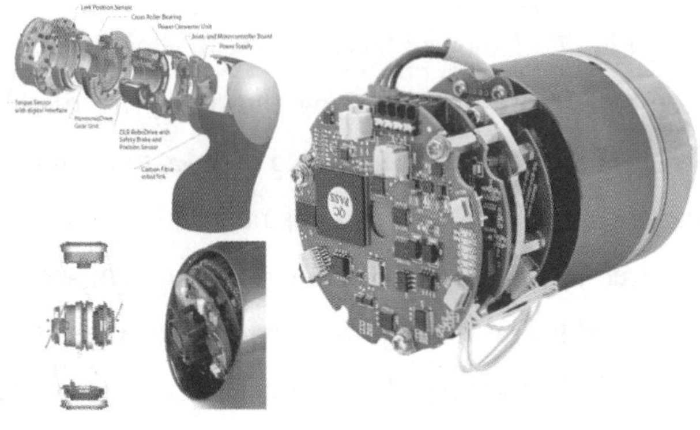

Fig.7-14 Modularity robot
图 7-14 模块化机器人

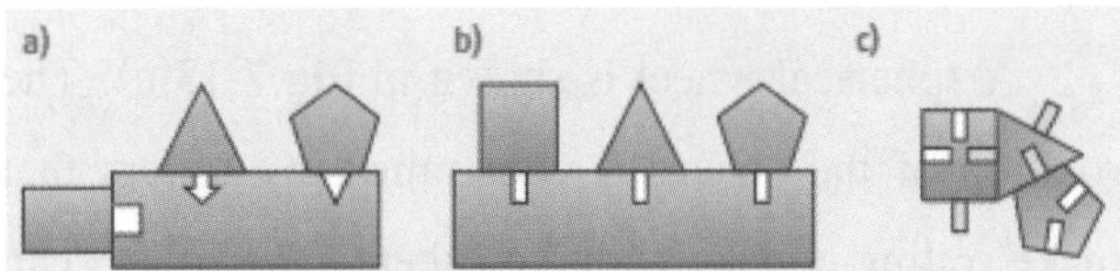

Fig.7-15 Assembly of components
图 7-15 组件装配

Modularity (Fig. 7-14) in design engineering refers to a compartmentalization of elements. Most often modularity in Complex systems occurs as a result of taking a complex system and dividing it into pieces in order to better understand simple module parallelization design. Modularity also facilitates the replacing of elements either for repair or upgrading new functionality. The alternative to a modular approach is an integrated approach where systems are

设计工程中的模块化（如图 7-14）指的是元件的划分。在复杂系统中，模块化通常是将一个复杂系统划分为多个部分以更好地理解简单的元件并行化设计工作的结果。模块化还有助于替换用于修复或升级新功能的元件。模块化方法的替代方法是一种集成方法，即系统作为一个整体来设计。虽然集成方法往往不那么容易修复、升级或重新配置，但它们对元件设计的限制较少，因此可以变得更优。集成方法可以专注于降低成本或获得更高的性能。

designed as a whole. While integrated approaches tend not to be as easy to repair, upgrade or reconfigure, they do have fewer constraints on element design and therefore can be made more optimal. Integrated approaches can focus on lowering cost or having higher performance.

在机械设备中，模块化或集成架构的选择对应用范围以及成本或性能有很大的影响。这种产品设计可以被用于机器人技术——可以被看作是工业或研究产品。产品模块化的架构可以分为三种子类型：插槽模块化、总线模块化和分段模块化。

In mechanical devices, the choice between modular or integrated architectures can have a large impact on the range of application as well as cost or performance. The product design literature can be looked at as encompassing robotics—which can be considered as an industrial or research product. The architecture of product modularity can be categorized into three subtypes: slot, bus, and sectional modularity:

插槽模块化：组件之间的每个接口都是不同类型的，因此产品中的各个组件不能互换。

Slot modularity: Each of the interfaces between components is of a different type from the others, so that the various components in the product cannot be interchanged.

总线模块化：有一个公共总线，其他物理组件通过相同类型的接口连接到这个公共总线。

Bus modularity: There is a common bus to which the other physical components connect via the same type of interface.

分段模块化：所有的接口都是相同类型的，并且没有一个单一的元件来附加所有其他组件，也就是说，没有基本组件。装配是通过相同的接口将组件相互连接起来的（如图7-15）。

Sectional modularity: All interfaces are of the same type and there is no single element to which all the other components attach, i. e., there is no base component. The assembly is built up by connecting the components to each other via identical interfaces shown in Fig. 7-15.

这样的分类为模块化机器人提供了一个很好的定义如下：

Such classifications provide a good definition of modular robots as follows:

If a complex robotic system adopts a slot-and bus-modularity design approach for its internal structure and architecture,not the external con-figuration, it can be called a modularly designed robotic system benefiting from design parallelization. Such a robot may have a unified and integrated configuration that cannot be changed from outside.

如果一个复杂的机器人系统在其内部结构和体系结构上采用插槽和总线模块化设计方法，而不是在外部构造上采用模块化设计方法，则可以称为得益于设计并行化的模块化设计机器人系统。这样的机器人有统一的、集成的配置，不能从外部改变。

If a robot adopts a bus- and sectional-modularity design approach for both internal structure and external configuration, it can be called a modular robot. The users can reconfigure the compartmentalization and interchange functional modules with some level of effort.

如果机器人的内部结构和外部结构采用总线模块化和分段模块化设计方法，则可称为模块化机器人。用户可以通过某种程度的努力重新配置划分和交换功能模块。

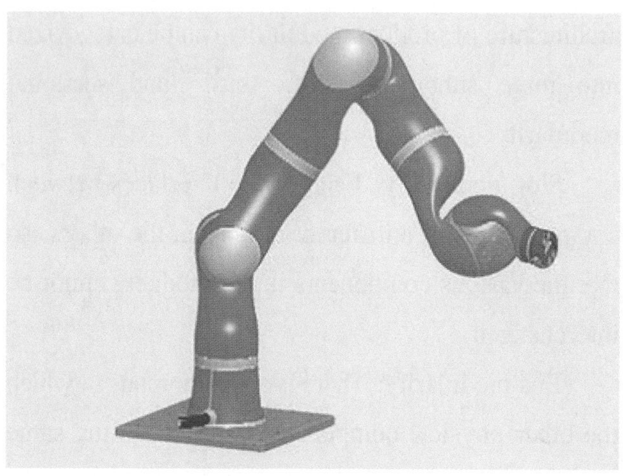

Fig.7-16 Redundancy robot
图 7-16 冗余机器人

Kinematic redundancy(Fig.7-16) occurs when a robotic manipulator has more degrees of freedom than those strictly required to execute a given task. This means that, in principle, no manipulator is inherently redundant; rather, there are certain tasks with respect to which it may become redundant. Because it is widely recognized that a fully general spatial task consists of following an end-effector motion trajectory requiring six degrees of freedom, a robot arm with seven or more joints is frequently

运动学冗余（图 7-16）设计用在机械臂的自由度大于执行给定任务所要求的自由度时。这意味着，从原则上讲，机械臂本身不是多余的；然而，对于某些任务，它可能变得多余。由于人们普遍认为，一个完全通用的空间任务需要六个自由度的末端执行器运动轨迹，具有七个或更多关节的机械臂通常被认为是一个固有冗余的机械臂的典型例子。然而，即使是像传统的六关节工业机械手（图 7-17）这样自由度较低的机械臂，对于特定的任务，如简单的末端执行器，也可能成为运动学上的冗余，从而在定位上不受约束。

considered as a typical example of an inherently redundant manipulator. However, even robot arms with fewer degrees of freedom (Fig. 7-17), like conventional six-joint industrial manipulators, may become kinematically redundant for specific tasks, such as simple end-effector positioning without constraints on the orientation.

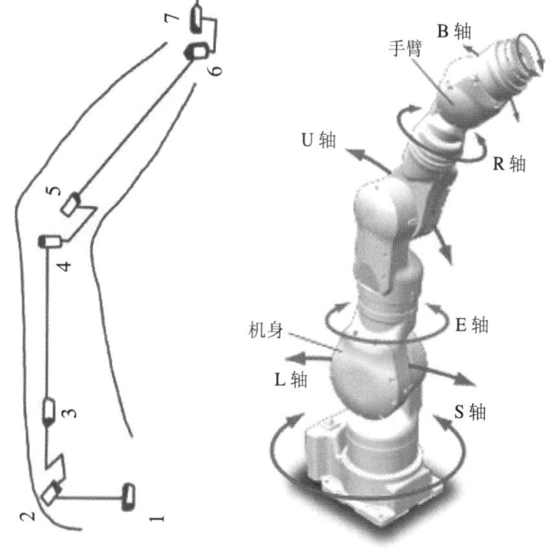

Fig.7-17 Six joint industrial robot
图 7-17 六关节工业机器人

在机械手的机械结构中引入运动学冗余的动机超出了传统工程设计中使用冗余的动机，即增加对故障的鲁棒性以提高可靠性（如冗余处理器或传感器）。事实上，赋予机器人机械手运动冗余主要是为了增加灵活性，就像人类的手臂一样。

The motivation for introducing kinematic redundancy in the mechanical structure of a manipulator goes beyond that for using redundancy in traditional engineering design, namely, increasing robustness to faults so as to improve reliability (e.g., redundant processors or sensors). In fact, endowing robotic manipulators with kinematic redundancy is mainly aimed at increasing dexterity, as is the case with a human arm.

早期机械手设计中，最小

The minimal-complexity approach that characterized

early manipulator designs had the objective of minimizing cost and maintenance. For example, this led to the development of selective compliance assembly robot arm （SCARA） robots for pick-and-place operations where products had been designed for assembly, using a single axis of insertion. However, giving a manipulator the minimum number of joints required to execute its task results in a serious limitation in real world applications where, in addition to the singularity problem, joint limits or workspace obstacles are present. All of these give rise to for bidden regions in the joint space that must be avoided during operation, thus requiring a carefully structured (and static) workspace where the motion of the manipulator can be planned in advance; this is the typical situation for work cells in traditional industrial applications.

On the other hand, the presence of additional degrees of freedom besides those strictly required to execute the task allows motions of the manipulator that do not displace the end effector (the so-called self-motions or internal motions); this implies that the same task at the end effector level can be executed in several ways at the joint level, giving the possibility of avoiding the forbidden regions and ultimately resulting in a more versatile mechanism. Such a feature is key to allowing operation in unstructured or dynamically varying environments that characterize advanced industrial applications and service robotics scenarios. That is where redundancy robots can have an advantage.

复杂性方法的目标是使成本和维护量最小化。例如，这导致了选择性顺应装配机器人手臂(SCARA)的发展，该机械手非常便于产品的拾取和放置操作，即使用单一插入轴。然而，在实际应用中，除了奇异性问题外，还存在关节限制或工作空间障碍，给机械手提供执行任务所需的最小关节数量会导致严重的限制。这些都导致了在操作过程中必须避免的关节在空间的禁止区域移动，因此需要一个精心构造的(静态)工作空间，在此可以预先规划机械手的运动；这是传统工业应用程序中工作单元的典型情况。

另一方面，除了执行任务所严格要求的自由度之外，还存在额外的自由度，从而允许操纵器的运动不移动末端执行器（所谓的自我运动或内部运动)；这意味着在末端执行器水平上的相同任务可以在关节水平上以多种方式执行，从而提供了避免禁止区域的可能性，并最终产生更通用的机制。这种特性是允许在非结构化或动态变化的环境中操作的关键，这些环境是高级工业应用程序和服务机器人场景的特征。这就是冗余机器人的优势所在。

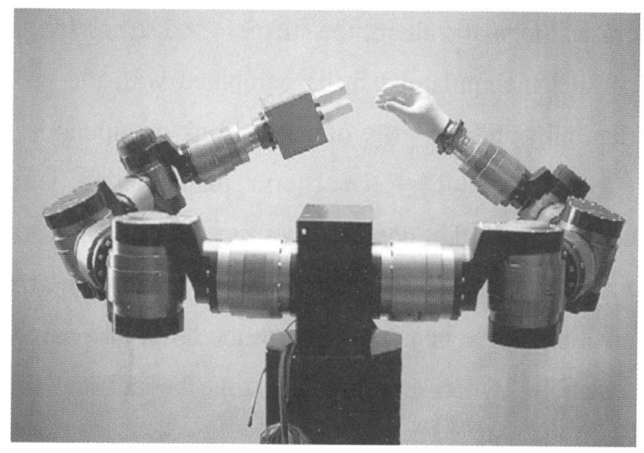

Fig.7-18 Dual arm robot
图 7-18 双臂机器人

There is an increasing trend of robots being moved into environments originally designed for human use. In industry, anthropomorphic robots (Fig.7-18) of human size are expected to replace human workers without major redesigns of the workplace. The ability to use human and robot workers inter changeably is thought to be the key to low-cost, flexible automation. As robot use in health care and domestic environments increases, so does the need for robots that are well adapted to these intrinsically human-centric environments. More advanced humanoid robots are expected to mimic human behaviors, act and manipulate objects in ways similar to humans.

There are several advantage of mechanical structure that motivate the use of dual arm setups:

Similarity to operator—The common motivation for using dual arms in teleoperation setups is that the operator is used to performing bimanual tasks, and having two arms at the slave site means that the operator's bimanual skills can be transferred to the remote site.

Flexibility and stiffness—By using two arms in a

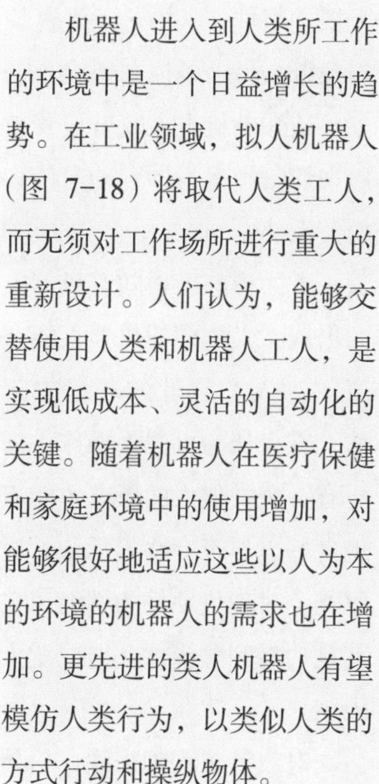
机器人进入到人类所工作的环境中是一个日益增长的趋势。在工业领域，拟人机器人（图 7-18）将取代人类工人，而无须对工作场所进行重大的重新设计。人们认为，能够交替使用人类和机器人工人，是实现低成本、灵活的自动化的关键。随着机器人在医疗保健和家庭环境中的使用增加，对能够很好地适应这些以人为本的环境的机器人的需求也在增加。更先进的类人机器人有望模仿人类行为，以类似人类的方式行动和操纵物体。

机械结构的几个优点促使双臂装置的应用：

与操作员的相似之处——在远程操作设置中使用双臂的共同动机是操作员习惯于执行双手操作的任务，而在从属位置有双臂意味着操作员的双手技能可以在远程位置使用。

closed kinematic chain, the stiffness and strength of a parallel manipulator can be combined with the task flexibility and dexterity of a serial link manipulator.

Manipulability—Dual arm manipulation is a contorl mode that controls two mechanisms to work for example a typical peg-in-hole task, with one arm positioning the peg and one arm the hole, or a screw assembly, where one arm controls the nut and the other the bolt .

Cognitive motivation—Humaniod dual arms have been set up to explore the relationship between humoid body interaction and conition.

Human form factor with the advent of robots that are expected to work in environments originally intended for humans, it is claimed that robot manipulators will need to have a human-like form to perform well, even though some acknowledge that this may place significant constraints on performance (especially power and robustness) with contemporary hardware.

The selection of a robot can be a significant problem for a design engineer, and the choice depends on a rage of factors, including the task to be performed. One of the earliest applications of robotics was within a foundry; such environments are hazardous to human operators due to noise, heat, and fumes from the process. This is a classic application of a robot being used to replace workers because of environmental hazards. Other tasks which suggest the use of robots include repetitive work cycles, the moving of difficult or hazardous materials, and requirements for multishift operation. Robot s that

柔性和刚度——通过在封闭运动链中使用双臂，可以使得并联机构的刚度和强度与串联机构的任务柔性和灵巧度相结合。

可操控性——双臂操作是同时控制两个机构来完成工作的控制方式，例如一个典型的钉入孔任务，用一条手臂定位钉，另一条手臂定位孔，或一个螺丝组件，其中一条手臂控制螺母，另一条控制螺栓。

认知动机类人双臂设置被用来探索类人身体互动与认知的关系。

人的形态因素随着机器人的出现，人们期望机器人能在原本为人类设计的环境中工作，有人认为机器人操作手需要有一个类似人的形态才能表现良好，尽管有些人承认，这可能会对现代硬件的性能（尤其是功耗和鲁棒性）造成重大限制。

对设计工程师来说，机器人的选择可能是一个重大问题，而选择取决于多种因素，包括要执行的任务。机器人技术最早的应用之一是在铸造厂；由于加工过程中产生的噪声、热量和烟雾，这种环境对人工操作人员是有害的。这是一个典型的应用，机器人被用

have been installed in manufacturing industry are normally employed in one of four application groups: materials handling, process operations, assembly, or inspection. Different tasks require different types of robots. In addition, the cost, operability and stability of the robot should also be considered.

来代替工人，因为环境危害。其他任务比如可以使用机器人进行重复的工作周期，移动困难或危险的材料，以及要求多班制操作，等等。在制造业中安装的机器人通常用于四个应用领域中的一个：物料搬运、工艺操作、装配或检验。不同的任务需要不同类型的机器人，此外，选择机器人的过程中，还要考虑价格、可操作性、稳定性等因素。

微信扫一扫

教学视频

Lecture 8-1

Robots Actuator(1)

8-1 工业机器人动力系统（1）

The ideal characteristics that are expected from a robot actuator are:

· A low inertia in order to increase the acceleration power of the joints;

· A high mechanical stiffness in order to minimize the deflection at tool level under the action of the load;

· A low operating speed V0: a few rad/s in rotary motion and a few tenths of cm/s in linear motion in order to obtain tool velocities from 1 to 5 m/s;

· A velocity range with continuous variation from−V0 to + V0;

· A high output force even at zero velocity, in order to provide an adequate holding torque;

· Low non-linearities (dry friction, backlash, etc.)

· The possibility to perform velocity and/or force control.

No actuator currently available fulfills all the conditions mentioned here above. The most suitable actuators are:

· Among the electric actuators:

– The step motors

机器人驱动系统的理想特性是：

·低惯量，以增加关节的加速度；

·具有较高的机械刚度，以最大限度地减小载荷的作用下的挠度；

·最低运行速度：在旋转运动中只有几 rad/s，在线性运动中只有零点几 cm/s，末端运动可以获得 1—5 米/秒的工具速度；

·从−V0 到+V0 连续变化的速度范围；

·高输出力，即使在零速度，以提供足够的保持扭矩；

·低非线性（干摩擦、间隙等）；

·执行速度和/或力控制的可能性。

目前没有执行机构能满足上面提到的所有条件。最合适的驱动器是：

·电动执行机构中：

– 步进电机

– 伺服电机

· 在液压执行机构中：

– 线性活塞和旋转活塞

– 带轴向活塞的旋转电机

· 气动执行机构中：

– 线性和旋转活塞

– Servo motors

· Among the hydraulic actuators:

– The linear and rotary pistons

– The rotary motors with axial pistons

· Among the pneumatic actuators:

– The linear and rotary pistons

Fig.8-1 Stepper motors
图 8-1 步进电机

步进电机（图 8-1 所示）的原理是直接将电子数字信号转换为增量信号：角度定位，步进电机驱动作为控制单元和电机的接口，接收不同频率的时钟脉冲，每个脉冲产生一个固定幅度的角位移，称为角度步进。这种关系在有限载荷和加/减速条件下得到验证，超过这些条件可能会产生定位误差。根据它的原理，步进电机是控制速度的，其主要优点是不需要闭环系统，因为旋转角度是成比例的时钟脉冲接收。

步进电机可以有三种不同

The principle of stepper motors (Fig.8-1) is to convert directly a electric digital signal into incremental angular positioning: the stepper motor driver, which plays the interfacing role between the control unit and the motor, receives clock pulses at a varying f-requency, each pulse generating an angular displacement of fixed amplitude called the angular step. This relationship remains verified under limited load and acceleration/deceleration conditions above which p-ositioning errors may occur. According to its principle, stepper motors are velocity controlled, and their main advantage is to avoid the need of a closed-loop servo system since the revolution angle is proportional to the number of clock pulses received.

Step motors can be of three different types:

- Motors with permanent magnet armature;
- Motors with variable reluctance;
- Hybrid motors.

The last ones gather the advantages of both former categories and are thus the more performing. They maintain a relatively low torque, also zero velocity, and their internal intertia is relatively high, so the power quality is low.

的类型：

- 带永磁电枢的电机；
- 可变磁阻电机；
- 混合动力电机。

最后一种集合了前两种的优点，因此性能更好。它们能够维持相对较低的转矩，也可以保持零速度，其内部惯性相对较高，因此功率质量比较低。

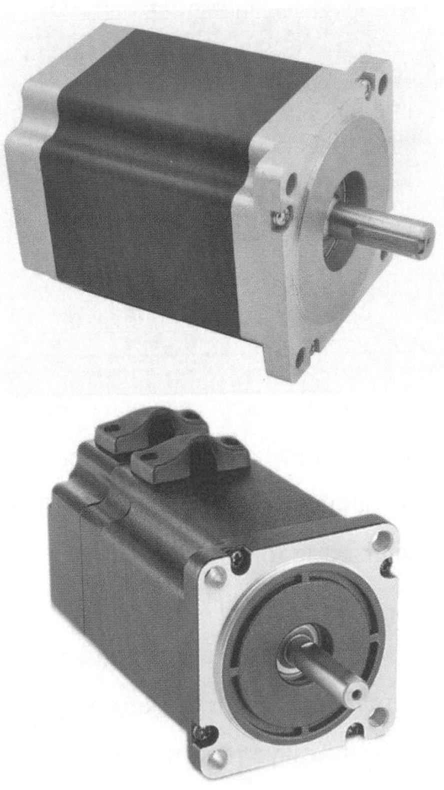

Fig. 8-2 Servo motor
图 8-2 伺服电机

Most robot manipulators use servomotors (Fig. 8-2) as a power supply. Servo motors are designed to accurately follow the desired position, velocity and torque which change frequently and sometimes abruptly. They have structures similar to ordinary

大多数机器人机械手使用伺服电机（图 8-2 所示）作为动力源。伺服电机的设计是为了精确地跟踪所需的位置、速度和扭矩，而这些位置、速度和扭矩经常变化，有时也会突然变化。它们的结构类似于普通的电动机，但具有低惯性和大转矩，可用于高加速度。用于机器人应用的典型伺服电机有永磁（PM）直流电机和直流无刷（B LDC）电机。

永磁直流电机具有转矩大、调速范围宽、矩速特性好、适应各种控制方法等优点，被广泛应用于伺服电机中。直流电机将电能转化为旋转或直线机械能。它有许多不同的类型和配置。成本最低的 PM 电机使用陶瓷（铁氧体）磁铁，机器人玩具和业余机器人经常使用这种类型的电机。采用稀土(钕铁硼)永磁定子可产生最大的扭矩和功率。无刷电机，也称为交流伺服电机或直流无刷电机，广泛应用于工业机器人。它们代替了电刷式直流电机中使用的石墨刷和铜棒换向器的磁性或光学传感器和电子开关电路，从而消除了换向部件的摩擦、火花和磨损。

无刷电机由于降低了电机的复杂性，通常具有较好的性能和较低的成本。然而，这些电机的控制器比电刷式电机控制器更复杂和昂贵。无刷电机的无源多极钕磁铁转子和线绕铁定子提供了良好的散热和良

electric motors, but with low inertia and large torque capably for high accelerations. Typical servomotors used for robotic applications are permanent magnet (PM) DC motors and brush less DC(B LDC) motors.

PM DC motors are widely used as a servomotor because of high torque, speed controllability over a w ide range, well behaved torque-speed characteristics, and adaptability to various types of control methods. The DC motor converts electrical energy into rotational or linear mechanical energy. It comes in many different types and configurations. The lowest-cost PM motors use ceramic (ferrite) magnets and robot toys and hobby robots often use this type of motor. A PM motor with a rare-earth (neodymium-iron-boron), NEO magnet stator produces the most torque and power for its size. Brushless motors, also called AC servomotors or brushless DC motors, are widely used in industrial robots. They substitute magnetic or optical sensors and electronic switching circuitry for the graphite brushes and copper bar commutator used in brush-type DC motors,thus eliminating the friction, sparking , and wear of commutating parts.

Brushless motors generally have good performance at low cost because of the decreased complexity of the motor. However, the controllers for these motors are more complex and expensive than brush-type motor controllers . A passive multipole neodymium magnet rotor and a wire-wound iron stator of a brushless motor provide good heat dissipation and excellent reliability. Linear brushless motors function like unrolled rotary motors. They typically have a long, heavy, multiple magnet passive stator and a short, lightweight, electronically commutated wire-wound forcer (slider).

The permanent-magnet, direct-current, brush-commutated motor is widely available and comes in many different types and configurations. The lowest-cost permanent-magnet motors use ceramic (ferrite) magnets. Robot toys and hobby robots often use this type of motor. Neodymium (NEO) magnet motors have the highest energy-product magnets, and in general produce the most torque and power for their size. Ironless rotor motors, often used in small robots, typically have copper wire conductors molded into epoxy or composite cup or disk rotor structures. The ad-vantages of these motors include low inductance, low friction, and no cogging torque. Disk armature motors have several advantages. They have short overall lengths , and because their rotors have many commutation segments, they produce a smooth output with low torque ripple. A disadvantage of ironless armature motors is that they have allow thermal capacity due to low mass and limited thermal paths to their case. As a result, when driven at high power levels they have rigid duty-cycle limitations or require forced-air cooling.

Hydraulic actuators Pistons are very simple and effective hydraulic systems which exist in either linear or rotary form. In their linear version, the displacement range is generally of a few centimeters or dizains of centimeters, while in the rotary version the angular displacement is limited to about 270 deg. Axial piston motors provide a continuous rotation but their technology is significantly more complicated.

好的可靠性。线性无刷电机的功能类似展开旋转电机。它们通常有一个长而重的多磁铁无源定子和一个短而轻的电子换向线绕转换器（滑块）。

永磁、直流、电刷换向电机应用广泛，并有许多不同的类型和配置。成本最低的永磁电机使用陶瓷（铁氧体）磁铁。机器人玩具和业余机器人经常使用这种类型的电机。钕（NEO）磁铁发动机具有最高的能源产品磁铁，一般来说，在其尺寸下产生最大的扭矩和功率。无铁转子电机，通常用于小型机器人，通常是铜线导体模铸成环氧树脂或复合杯或磁盘转子结构。这些电机的优点包括低电感、低摩擦、无齿槽转矩。圆盘电枢电机有几个优点。它们的总长度很短，因为内部转子有许多换相段，能够产生平稳的输出与低转矩波纹。无铁电枢电机的一个缺点是热容量、低质量和有限的热会传递到它们的外壳。因此，当驱动在高功率水平，它们有严格的脉宽周期限制或需要强制空气冷却。

液压电机是非常简单和有效的液压系统，存在于线性或旋转形式。在它们的线性版本中，位移范围通常是几厘米或几十厘米，而在旋转版本中，角位移被限制在 270 度左右。轴向活塞电机提供了一个连续的旋转，但它们的技术是很复杂的。

它们的控制是通过一个作用在分配器上的小电动机或使用伺服阀来实现的，伺服阀提供了控制单元和执行器本身之间的接口。液压执行机构的主要优点是，低内部惯性和高加速度功率，还具有高机械刚度的特点。

然而，它们表现出大量不同来源的非线性，这使得相关反馈系统难以稳定。普通活塞可以维持在零速度，而不出重大问题。但对于轴向活塞，则会呈现重要的扭矩和速度振荡而接近于零速度。

气动执行机构类似于液压执行机构。气动执行器将能量(以压缩空气的形式)转换为机械运动，可以是线性的，也可以是旋转的。气动执行机构主要存在于简单的机械手中。典型地，它们提供不受控制的运动之间的机械限制停止。这些执行机构提供良好的性能点对点的运动。它们易于控制，成本低。与电动马达相比，气动马达有几个优点。它们在危险环境中相对安全；它们受环境

Their control is achieved either through a small electric motor acting on a distributor or using a servovalve which provide the interface between the control unit and the actuator itself. The main advantages of hydraulic actuators are, as already mentioned, their low internal inertia and high acceleration power. They are also characterized by a high mechanical stiffness.

However, they exhibit a large number of nonlinearities of different origins which make the associated feedback system difficult to stabilize. The ordinary pistons do not raise significant problems when maintained at zero speed, but this is not the case of axial piston actuators which present important torque and velocity oscillations near zero speed.

Pneumatic actuators are similar to hydraulic actuators. Pneumatic actuators convert energy (in the form of compressed air) into mechanical motion , which may be linear or rotary. Pneumatic actuators are primarily found in simple manipulators. Typically they provide uncontrolled motion between mechanical limit stops .These actuators provide good performance in point-to-point motion. They are simple to control and are low in cost. Pneumatic motors have several advantages over electric motors. They are relatively safe in the explosive environment. They are also less affected by ambient temperature and humidity than electric motors. Although a few small actuators may be run with typical factory air supplies, extensive use of pneumatic-actuated robots requires the purchase and installation of a costly dedicated compressed -air source.Pneumatic actuators have low energy efficiency. Pneumatic systems consist of pneumatic genera-tor, pneumatic valves, pneumatic actuator and pipes.

A pneumatic generator produces compressed air using an air compressor. Pneumatic valves are used to control the pressure, flow rate and direction . The mechanical motion produced may be linear for pneumatic cylinders or rotary for pneumatic motors . Pneumatic actuators are not used for applications requiring large forces or torques since they produce less power than hydraulic actuators or electric actuators. However, they are used in robot hands or artificial muscles, which require high power-to-weight ratios. Pneumatic artificial muscles are contractile or extensional devices operated by pressurized air filling a pneumatic bladder. They are usually grouped in the agonist and antagonist pairs. Furthermore, they can be used for medical robots since they are not affected by magnetic field, and for robots in explosive environments because there is no electrical arcing that exists in electromagnetic actuators.

Motion-control systems

As described earlier, the most extensive robot actuator is servomotors. In order to achieve this motion-control, the relationship between the mechanical aspects of the robotic system and its actuators needs to be fully understood. Even with the best control system and algorithms available, it will not perform to specification if the load cannot be accelerated or decelerated to the correct speed within the required time and if that speed cannot be held to the required accuracy.

A motion-control system consists of several elements whose characteristics must be carefully

温度和湿度的影响也较小。虽然一些小型执行机构可能与典型的工厂空气供应一起运行，但广泛使用气动执行机构机器人需要购买和安装一个昂贵的专用压缩空气源。气动执行机构的效率很低。气动系统由气动发生器、气动阀门、气动执行机构和管路组成。

气源使用空气压缩机产生压缩空气。气动阀用于控制压力、流量和方向。所产生的机械运动对于气动气缸可以是线性的，对于气动马达可以是旋转的。气动执行机构不用于需要大的力或扭矩的应用，因为它们产生的功率比液压执行机构或电动执行机构小。然而，它们被用于机器人的手或人造肌肉，这需要高的功率-重量比。气动人造肌肉是一种收缩或伸展装置，由充气的加压空气控制。它们通常分为激动剂对和拮抗剂对。此外，它们可以用于医疗机器人，因为它们不受磁场的影响；危险环境中，因为没有电弧存在的电磁执行器。

运动控制系统

如前所述，最广泛的机器人执行机构是伺服电机。为了实现这种运动控制，需要充分理解机器人系统的机械方面和

determined in order to optimise the performance of the complete system. A motion control system consist s of five elements:

The controller that implements the main control algorithms (normally either speed or position control) and provides the interface between the motion-control system and the main control system and/or the user.

The encoders and transducers required to provide feedback of the load's position and speed to the controller.

The motor controller, and motor. In most cases, these can be considered to be an integral package, as the operation and characteristics of the motor being totally dependent on its control package. Depending on the motor type, position and current feedback may be required.

The transmission system. This takes the motor output and under takes the required speed changes and, if required, a rotary-to-linear translation.

The load. The driven elements greatly influence the operation of the complete system. It should be noted that a number of parameters, including inertia, external loads, and friction, may vary as a function of time, and need to be fully determined at the start of the design process.

The key to successful implementation of a drive system is full identification of the applications needs and hence its requirements. In order to select the correct system for an application, a number of activities,ranging over all aspects of the appli-cation, have to be undertaken. The key stages of the process can be identified as follows:

执行机构之间的关系。即使用最好的控制系统和算法，如果负载不能在要求的时间内加速或减速到正确的速度，如果速度不能保持到要求的精度，系统整体也将不会达到要求。

一个运动控制系统由几个元件组成。必须仔细确定其特性，以优化整个系统的性能。运动控制系统由五个要素组成：

实现主控制算法的控制器（通常是速度控制或位置控制），并提供运动控制系统和主控制系统和/或用户之间的接口。

编码器和传感器需要向控制器反馈负载的位置和速度。

电机控制器、电机。在大多数情况下，这些可以被认为是一个完整的系统，因为电机的操作和特性完全依赖于它的控制器。这取决于电机类型、位置和电流反馈可能需要有关参数。

传输系统。这取决于电机输出和所需的速度变化，如果需要，还要实现旋转到线性的转化。

负载。驱动元件对整个系统的运行影响很大。应该注意的是，许多参数，包括惯性、外部载荷和摩擦，可能随时间变化，需要在设计过程的开始

Collection of the data. The key to satisfactory selection and commissioning of a motor drive system is the collection of all the relevant data before starting the sizing and selection process. The information obtained will mostly relate to the system's operation but may also include commercial considerations.

Sizing of the system. The size of the various drive components can be determined from the data collected earlier.

Identification of the system to be used. Once the size of the various elements and the application requirement s are known, the identity of the various elements can be indicated. At this stage, the types of the motor,feedback transducer, and controller required can be finalised.

Selection of the components. Using the acquired knowledge, the selection process can be started. If the items cannot be supplied and the selection revised, or the specification of a component is changed, the effect on the complete system must be considered.

Verification. Prior to procuring the components, a complete check must be made to ensure that the system fits together in the space allocated by the other members of the design team.

Testing. Theoretically, if all the above steps have been correctly followed , there should be no problems. But this is not always the case in the real world, commissioning modification may be required. If this is required care must be taken to ensure that the performance of the system is not degraded.

One of the main design decisions that is required is the selection of the correct motor technology. With

就完全确定。

成功搭建一套驱动系统的关键，是充分了解应用系统的需要，从而确定系统的要求。为了给应用程序选择正确的系统元素，必须进行一系列的活动，范围涉及应用程序的各个方面。该过程的关键阶段可确定如下：

收集数据。对电机驱动系统进行适当的选择和调试的关键是在开始筛选和选择过程之前收集所有相关数据。所获得的资料将主要与系统的运作有关，但也可能包括商业考虑。

系统的规模调整。各种驱动器组件的大小可以从之前收集的数据中确定。

要使用的系统标识。一旦知道了各种元素的大小和应用程序需求，就可以指出各种元素的标识。在这个阶段，电机、反馈传感器和控制器的需求类型可以最终确定。

组件的选择。利用已获得的知识，可以开始选择过程。如果项目不能提供，选择被修改或者组件的规格被改变，必须考虑对整个系统的影响。

验证。在采购组件之前，必须进行一次完整的检查，以确保该系统与设计团队其他成员分配的空间相匹配。

测试。从理论上讲，如果以上所有步骤都被正确地验证了，应该没有问题。但在现实中情况并不总是这样，可能需要调试修改。如果需要这样做，则必须小心确保系统的性能不会降低。

所需要的主要设计决策之一是选择正确的电机技术。随着该领域的快速发展，可供选择的方案越来越多；每种选择都有优缺点。在考虑整个系统时，电机决定了驱动的特性，从而决定了功率转换和控制要求。

考虑电机的扭矩因素

为了实现运动控制系统的完美运行，系统内的所有组件都必须经过精心选择。如果选择不正确，无论是在类型或大小的电机和/或驱动器的任何轴，整个系统的性能将受到影响。应该认识到，型号过大与型号过小一样不可行；这个系统都可能打折扣，而且会增加更多成本。在最广泛的意义上，选择电机驱动器被认为需要系统地收集有关轴的数据，并对其进行随后的分析。

一般来说，电机通过传动系统中的某种传动系统来驱动负载，虽然电动机总是旋转，但负载也可以旋转或进行平移运动。完整的套件可能还包括

the rapid development in this field, number of options are available; each option will have benefits and disadvantages. In the consideration of the complete system the motor determines the characteristic of the drive, and hence determines the power converter and control requirements.

Torque considerations of motors

To achieve satisfactory operation of any motion control system, all the components within the system must be carefully selected.

If an incorrect selection is made, either in the type or the size of the motor or drive for any axis, the performance of the whole system will be compromised. It should be realised that over-sizing a system is as bad as under sizing; the system may not physically fit and will certainly cost more.

In the broadest sense, the selection of a motor-drive can be considered to require the systematic collection of data regarding the axis, and its subsequent analysis.

In general, a motor drives the load through some form of transmission system in a drive system. Although the motor is always rotating, the load also be retated or translated. The complete package will probably also include a speed changing system, such as a gearbox or belt drive.

The first step to the successful sizing and selection of a system is the collating of the information about the system and its application. Apart from the electrical and mechanical aspects this must also include details of the operating environment; for if these details are not considered at an early stage, the system which is selected may not be suitable for the application.

一个变速系统，如变速箱或皮带传动。

成功确定和选择系统的第一步是整理关于系统及其应用程序的信息。除了电气和机械方面，还必须包括操作环境的细节；因为如不及早考虑这些详情，所选用的系统都可能不适合。

微信扫一扫

教学视频

Lecture 8-2

Robots Actuator(2)

8-2 工业机器人动力系统（2）

电磁干扰（EMI）可不同程度地影响各种类型的电气和电子设备；近期的规范显示出在设计阶段干预并消除这些潜在问题的重要性。系统既能发出电磁干扰，又易受电磁干扰影响，因此在设计和使用系统时必须小心谨慎。在评估设备兼容性时，必须在一个非常宽的频率范围内考虑设备的发散和敏感性，通常是直流（0 赫兹）到 110 兆赫兹。因此，电磁兼容性对系统的设计和应用有相当大的影响。电磁辐射最明显的来源是用于电机控制器和任何相关开关装置的功率转换器。随着微处理器控制器的使用越来越多，任何干扰都可能产生严重的安全隐患。一般来说，进出设备的主要干扰途径是电缆或直接辐射。长度超过 1 米的电缆是已发生区。接地和屏蔽的使用不是优先考虑的问题，因为必须仔细设计面板的布局。虽然实施这些措施的

Electromagnetic interference (EMI) can affect all types of electrical and electronic equipment to varying degrees; such interference has increased in importance because of recent legislation, which recognises the importance of removing these potential problem areas at the design stage. It should be recognised that a system can both emit and be susceptible to EMI, hence care must be taken in the design and use of a system. In an assessment of equipment for compatibility, the emissions and the susceptibility of the equipment must be considered over a very broad frequency range, normally d.c. (0 Hz) to 110 MHz. In consequence, electromagnetic compatibility has a considerable in fluence over the design and application of a system. The most obvious sources of electromagnetic radiation are the power converters which are used in motor controllers and any associated switch gear. With the increasing use of microprocessor-based controllers, any interference can have serious safety implications. In general, the main route of interference into or out of a piece of equipment is via the cabling or direct radiation. A cable which is longer than one metre should be considered to be a problem area. The use of the grounding and shielding of cab les m us t be a high priority, as must careful design of the panel layout. While these measures cost little to implement, failure to do so will be costly. Radio- frequency interference (RFI) is electromagnetic emission in the range 150 kHz to 110 MHz which are of sufficient strength to be capable of interfering with any form of telecommunications; the main sources of RFI are power electronics, relays , and motor commutators. Recent legislation has placed strict limits on RFI emissions, and these limits should be met before a piece of equipment can be placed in service.

Due to problems of interference between systems, a range of standards detail the amount of RFI which is permitted to be emitted and , perhaps more importantly a systems susceptibility. The suppress ion of RFI requires considerable thought and can only be achieved by the use of filters and good design practice. The screening of cab les will not affect the RFI being transmitted around a system. It should be noted that these directive state that screened cables are not an acceptable form of RFI emission prevention. The only acceptable form of suppression is by the use of in-line filters in all the supply lines, including the earth return. The design of suitable filtering should be carried out in partnership with the system user, its supplier, and the electricity supply authority.

Wiring considerations of motors

While the installation of motor power cables needs to be considered for any EMI and RFI effects, other factors must also be considered if the separation of the motor and drive is excessive . The pulses from

成本很低，但如果不预置，将会付出更大代价。射频干扰(RFI) 是指在 150 kHz 到 110 MHz 范围内的电磁发射，其强度足以干扰任何形式的通信；RFI 的主要来源是电力电子、继电器和电机换向器。最近的规范对 RFI 排放设置了严格的限制，在设备投入使用之前，必须满足这些限制。

由于系统间的干扰问题，一系列标准详细规定了允许发射的 RFI 的数量，也许更重要的是系统灵敏度。抑制 RFI 离子需要深思熟虑，只能通过过滤器的使用和良好的设计实践来实现。屏蔽不会影响在系统周围传输的 RFI。应该指出的是，这些指令表明屏蔽电缆不是 RFI 发射预防的可接受的形式。唯一可接受的抑制形式是在所有供电线路（包括接地回路）上使用在线过滤器。适当过滤的设计方案应与系统用户、其供应商和电力供应当局合作制定。

电机接线注意事项

虽然已经考虑了安装电机电力电缆的 EM I 和 RFI 影响，但如果电机和驱动器的距离太远，还必须考虑其他因素。来自脉冲宽度调制驱动器的脉冲将作为一个脉冲作用在电机电缆上，并会由于电机和电缆之间的阻抗不匹配而发生反射。如果电缆的长度过大，反射脉冲将与其他脉冲结合，使电机电压增加到标准电源电压的两倍以上。从传输线理论的角度来看，反射波理论是很好理解的。应当考虑高电压尖峰的净效应，电机的绝缘将打破。在选择电机时，应考虑以下几点：

如果要将电机改造成绝缘规格未知的电机，这个问题只能通过与原电机供应商协商来解决，并且可能需要用具有增强绝缘性能的电机。

在驱动系统的供电系统设计中，确保系统接地是很重要的。良好的接地系统是有必要的：

通过将触碰电压限制在安全值，为故障电流提供低阻通道，使电路保护装置快速运行，切断电源，以确保操作人员和其他人员的安全。接地线路的电阻必须足够低，以使接地端子上的电位上升和与之连接的任何金属制品都没有危险。

通过提供接地，限制 EMI 和 RFI 的影响。

保证供电网络的正常运行，保证良好的电能质量。

一个完整的接地系统的实际设计很复杂，应参考相关的

a pulse width modulated drive will act as an impulse on the motor cables, and reflection will occur due to the mismatch in impedance between the motor and the cable. If the length of the cable is excessive, the reflected pulse will combine with other pulses to increase the motor voltage to over twice the nominal supply voltage. This reflected-wave theory is well under stood in terms of transmission line theory. The net effect of this high-voltage spike is the possibility that the motor' s insulation will be broken. In the selection of a motor, the following points should be considered:

If the drive is to be retrofit ted to a motor with unknown insulation specifications, this problem can only be resolved by consultation with the supplier of the original motor, and it may require replacement of the motor by a motor with enhanced insulation capabilities.

In the design of the electrical supply system to a drive system, it is important to ensure that the system is fully and correctly earthed. A good earthing system is required to:

Ensure safety of operators and other personnel by limiting touch voltages to safe values, by provide a low resistance path for fault current so that the circuit protective devices operate rapidly to disconnect the supply. The resistance of the earth path must be low enough so that the potential rise on the earth terminal and any metalwork connected to it is not hazardous.

Limit the amount of EMI and RFI, by providing a noise-free ground.

Ensuring correct operation of the electricity supply network and ensure good power supply quality.

The actual design of a complete earthing system is complex, and reference should be made to the relevant national standards.

In the construction of a drive system, bonding is applied to all accessible metalwork-whether associated with the electrical installation (known as exposed-metalwork) or not (extraneous- metal work) - is connected to the system earth. The bonding must be installed so that the removal of a bond for maintenance of the equipment does not break the connection to any other bond.

The provision of a good earth is fundamental to the prevention of EMI and RFI problems. It is common practice to use a single point or star earthing system to avoid the problems of common mode impedance coupling. However, care needs to exercised when shielded or screened cables are used, as loops may inadvertently be formed, which will provide a path for any noise current.

Supply considerations of motors

While the quality of public-utility supplies is normally controlled within tight specifications, considerable voltage fluctuations may have to be accommodated in a particular application. In cases were the drive system is used on sites with local generation (for example, on oil rigs and ships), considerable care needs to be taken in the specification of the voltage limits. Since the peak speed of a motor is dependent on the supply voltage, consideration needs to be given to what happens

国家标准。

在驱动系统的构造中，将所有可触及的金属件接地——无论与电气安装有关（称为裸露金属件）与否（无关的金属件）。连接的安装必须保证为维护设备而拆除连接不会破坏与其他连接。

提供良好的接地是预防EMI和RFI问题的基础。通常采用单点或星形接地系统，以避免共模阻抗耦合问题。然而，当使用屏蔽或屏蔽电缆时需要小心，因为可能无意中形成回路，这将为噪声电流提供路径。

电机供电注意事项

虽然普通供电的质量通常控制在严格的规格内，但在特定的环境中，可能需要适应相当大的电压波动。如果驱动系统用于有本地发电的现场（例如，在石油钻机和船舶上），则需要在电压限制的规范中格外小心。由于电动机的峰值速度取决于供电电压，因此需要考虑在低压期间发生的情况。一般情况下，驱动器通常运行在最高速度的80%的名义供电电压下。如果系统是由不稳定的供应源提供的，则必须非常小心，以确保驱动器、控制器和负载都受到保护，不受损害；这个问题在引入微处理器系统后尤其严重，如果没有正确配置，系统可能会在没有警告的情况下锁定或复位，从而可能导致严重的故障。

在实践中，由于下列干扰，电源电压可能偏离一个完美的正弦波。

过电压。在相当多的周期内，电压幅值实质上高于其标称值。这可能会导致系统负荷突然下降，从而导致供应迅速上升。

欠压或低电压。在相当多的周期内，电压大大低于其标称值。电压不足可由负载突然增加引起，例如机床或感应电动机启动。

停电或故障。在一段时间内，供应峰值会降至零，可能是几个周期，也可能是一段较长的时间。

电压峰值。这些是叠加在正常供电波形和非重复性。尖峰可以是差分模或共模。功率因数校正电容器、电力线或附近电机的快速开关，偶尔会造成较大的电压尖峰。

during a period of low voltage. As a guideline, drives are normally sized so that they can run at peak speed at eighty percent of the nominal supply voltage. If a system is fed from a vulnerable supply, considerable care will have to be taken to ensure that the drive, its controller,and the load are all protected from damage; this problem is particularly acute with the introduction of microprocessor systems, which may lock-up or reset without warning if they are not properly configured, leading to a possibly catastrophic situation.

In practice, the supply voltage can deviate from a perfect sine wave due to the following disturbances.

Overvoltage. The voltage magnitude is substantially higher than its nominal value for a significant number of cycles.This can be caused sudden decreases in the system load , thus causing the supply to rise rapidly.

Undervoltage or brownout. The voltage is substantially lower than its nominal value for a significant number of cycles. Under voltages can be caused by a sudden increase in load, for example a machine tool or induction motor starting.

Blackout or outage. The supply collapses to zero for a period of time that can range from a few cycles to an extended period of time.

Voltage spikes. These are super imposed on the normal supply waveform and are non-repetitive. A spike can be either differential- mode or a common-mode. Occasional large voltage spikes can be caused by rapid switching of power factor correction capacitors, power lines or motors in the vicinity.

Chopped voltage waveform. This refers to a

repetitive chopping of the waveform and associated ringing. Chopping of the voltage can be caused by ac-to-dc line frequency thyristor converters.

截断的电压波形。这指的是波形和相关的振幅重复。电压的截断可以由交流到直流线路频率晶闸管转换器造成。

Harmonics.A distorted voltage waveform contains harmonic voltage components at harmonic frequencies (typically low order multiples of the line frequency). These harmonics exist on a sustained basis. Harmonics can be produced by a variety of sources including magnetic saturation of transformers or harmonic currents injected by power electronic loads.

谐波。畸变的电压波形包含谐波频率（通常是线路频率的低阶倍数）的谐波电压分量。这些谐波是持续存在的。谐波可以由多种来源产生，包括变压器的磁饱和或电力电子负载注入的谐波电流。

Electromagnetic interference.This refers to high-frequency noise, which may be conducted on the power line or radiated from its source.

电磁干扰。这是指高频噪声，可能在电源线上传导，也可能从电源辐射而来。

The effect of power line disturbances on drive systems depends on a number of variables including the type and magnitude of the disturbance, the type of equipment and how well it is designed and constructed, and finally the power conditioning equipment fitted to the system or the individual drive.

电力线干扰对驱动系统的影响取决于许多变量，包括干扰的类型和大小，设备的类型和它的设计和构造的好坏，最后是适合系统或单个驱动器的电力调节设备。

Sustained under-and over-voltages will cause equipment to trip out, which is both highly undesirable and with a high degree of risk in certain applications. Large voltage spikes may cause a hardware (particularly in power semiconductors) failure in the equipment . Manufacturers of critical equipment often provide a certain degree of protection by including surge arrestors or snubbers in their designs. However, spikes of very large magnitude in combination with a higher frequency may result in a stress-related hardware failure , even if normal protection standards are maintained. A chopped voltage and harmonics have the potential to interfere with a drive system if it is not designed to be

持续的电压不足和过电压会导致设备跳闸，应当避免出现这种情况，在某些应用中有很高的风险。较大的电压峰值可能会导致设备中的硬件（特别是功率半导体）故障。关键设备的制造商通常通过在设计中加入避雷器或缓冲器来提供一定程度的保护。然而，即使保持了正常的保护标准，非常大的峰值加上更高的频率也可能导致与压力相关的硬件故障。如果没有设计预防措施，截断的电压和谐波有可能干扰驱动器系统。

功率调节提供了一种有效的方法来抑制部分或所有的电气干扰，而不是停电和频率。提供电源调节的典型方法包括：

提供线路模式电压尖峰保护的金属氧化物压敏电阻，电磁干扰滤波器，有助于防止截断波形对设备的影响，以及防止设备将高频噪声传导到电源中。

具有静电屏蔽的隔离变压器，它不仅提供电流隔离，而且提供对电压尖峰的保护。

谐振变压器，提供电压调节和尖峰滤波。

所有的电力电子转换器（包括那些用来保护临界负载的转换器）都可以通过扭曲供电波形增加供电线路的干扰，为了说明电力电子负载输入电流中的电流谐波问题。由于电源的有限内阻抗，在与其他负载共同耦合点的电压波形会发生畸变，这可能会造成额外的故障。除波形畸变外，谐波电流引起的其他问题还包括：公用配电和设备中的过热和过电压（由于谐振条件），计量环错误，公用继电器故障，通信和控制信号的干扰。

减小这种影响的一种方法是过滤电力电子负载产生的谐

immune from such effects.

Power conditioning provides an effective way of suppressing some or all of the electrical disturbances other than the power outages and frequency. Typical methods of providing power conditioning include:

Metal-oxide varistors, which provide protection against line-mode voltage spikes, Electromagnetic interference filters, which help to prevent the effect of the chopped waveforms on the equipment as well as to prevent the equipment from conducting high-frequency noise into supply.

Isolation transformers with electrostatic shields, which not only provide galvanic isolation, but also provide protection against voltage spikes.

Ferroresonant transformers,which provide voltage regulation as well as line spike filtering.

All power electronic converters (including those used to protect critical loads) can add to the supply line disturbances by distorting the supply waveform. To illustrate the problems due to current harmonies in the input current of a power electronic load. Due to the finite internal impedance of the supply source, the voltage waveform at the point of common coupling to other loads will become distorted , which may cause additional malfunctions. In addition to the waveform distortion, other problems due to the harmonic currents include: additional heating and over-voltages (due to resonance conditions) in the utility distribution and equipment, errors in mete ring and malfunction of utility relays and interference with communications and control signal.

One approach to minimise this impact is to filter

the harmonic currents and the EM I produced by the power electronic loads. An alternative, despite a small increase in the initial cost, is to design the power electronic equipment such that the harmonic currents and the EMI are prevented or minimised from being generated in the first place.

波电流和电磁。另一种选择，尽管在初始成本上略有增加，是设计电力电子设备，使谐波电流和电磁干扰在一开始就得到控制。

A significant proportion of drive systems have to operate in relatively poor environments. The first line in this protection is the provision of a suitable protective enclosure. Two basic classes exist for non-hazardous areas and hazardous areas. An enclosure for non-hazardous areas is classified by the use of an IP code number, which indicates the degree of protection from ingress of solid objects including personnel contact, dust, and liquid.

很大一部分驱动系统必须在相对较差的环境中运行。这种保护的第一道线是提供适当的保护外壳。分为无危险区域和危险区域两种基本类别。非危险区域的外壳通过IP编码的使用进行分类，IP 编码表示防止物体（包括人员接触、灰尘和液体）进入的防护程度。

The IP (internal protection) rating system specify designers a motor, enclosure, or other components, with a specified degree of protection from dust, water, and impact. The two numerals can be used to specify the protection afforded to a component. In a number of cases a third numeral is added, which defines the impact rating of the enclosure. The definitions of the IP classifications are given in Table. If a drive must operate in a hazardous environment, where an explosive gas /air mixture is present. It is recommended that the designers of systems for this type of environment consult the relevant specialist agencies, e.g., Health and Safety Agency. These general application problems can never be solved by one specific formula; rather, the requirements of the various equipment must be recognised, and an optimum system should be

IP 分类的定义见表。如果驱动器必须在存在爆炸性气体/空气混合物的危险环境中工作。建议这类环境的系统设计者咨询相关专业机构，如健康和安全机构。这些普遍的应用问题永远不能用一个特定的公式来解决；相反，必须认识到各种设备的要求，并通过仔细注意细节来选择最佳的系统。例如，一个保护 IP5,4 的系统可以防止灰尘和水喷，尽管允许有限的泄漏。但是，如果该系统将用于危险环境，例如爆炸性环境，就适用于这些规定了。

IP等级系统向设计者列出了电机、外壳及其他组件，并指定了防尘、防水和防冲击等级这两个数字可用于指定为组件提供保护。在大多数情况下，会添加第三个数据，用于定义外壳冲击等级。该表给出了 IP 认证的定义。

selected by careful attention to detail. For example, a system protected to IP5,4 is protected against dust and water being sprayed in all direction, though limited leakage is permitted . If however the system is to be used in a hazardous environment, for example an explosive atmosphere, different regulations apply.

	First number	**Second number**
	Protection against solid objects	**Protection against liquids**
0	No protection	No protection
1	Objects up to 50.00 mm	Protection against falling drops of water
2	Objects up to 12.50 mm	Dripping water up 15 degrees from the vertical
3	Objects up to 2.50 mm	Direct sprays up 60 degrees from the vertical
4	Objects up to 1.00 mm	Water splashed in all direction; limited ingress is permitted
5	Protection against dust; limited ingress is permitted, but no harmful deposits	Protection from low pressure jets of water in all directions.
6	Totally protected against dust	Protection from powerful water jets in all direction
7		Protection from immersion in water up to a depth between 15 cm and 1 m
8		Immersion under water to a specified depth, typically 3 m.

微信扫一扫

教学视频

Lecture 9-1
Perception System of Industrial Robots(1)

9-1 工业机器人的感知系统（1）

Course import:

How can an industrial robot perceive its own state and the position of a target in its work, so that the end-effector can accurately walk to the designated position, and how can it perceive the shape and color of external objects, whether there are obstacles around the object and other environmental information.

1. The Perception System of Industrial Robots

The perception system of an industrial robot is similar to that of human, which can perceive image information, frequency information, pressure information, odor information, temperature information, balance information, and other. The core component of the robot's perception system is the sensor, which can transform various physical information into electrical signals through physical and chemical means. Generally, sensors can be divided into internal sensors and external sensors according to the purpose and scope of application of the sensors on the robot.

2. Internal Sensors

The sensors are used to detect the robot's own state (such as the angle between the arms, the position, speed, acceleration, etc. of the robot during movement)

课程导入：

工业机器人在工作中如何能感知自身的状态和目标物的位置，以便让末端执行器准确地移动到指定的位置，以及如何能感知外界物体的形状和颜色，物体周围是否有障碍物和其他的环境信息。

1．工业机器人的感知系统

工业机器人的感知系统类似于人的感知系统，能够感知图像信息、频率信息、压力信息、气味信息、温度信息、平衡信息以及其他信息。机器人的感知系统的核心部件是传感器，它能够通过物理的和化学的手段，将各种物理信息转变为电信号。一般情况下，根据传感器在机器人上应用目的和使用范围的不同可将传感器分为内部传感器和外部传感器。

2．内部传感器

用于检测机器人的自身状

态（如手臂间角度，机器人运动过程中的位置、速度、加速度等）的传感器是内部传感器。

2.1 内部传感器的功能

内部传感器的主要功能如下：

（1）规定位置，规定角度的检测

检测预先规定的位置或角度，可以用开/关两个状态，用于检测机器人的起始原点、极限位置或确定位置。

（2）位置、角度测量

测量机器人关节线位移和角位移的传感器是机器人位置反馈控制中必不可少的元件，包括电位器、旋转变压器、编码器。

（3）速度、角速度测量

最通用的速度、角度传感器是测速发电机或转速传感器、比率发电机。其中，测量角速度的测速发电机可按其构造分为直流测速发电机和感应式交流测速发电机。

（4）加速度测量

加速度传感器包括应变片式加速度传感器、伺服加速度传感器、压电感应加速度传感器及其他类型加速度传感器。

2.2 位置传感器

位置传感器包括位置和角度检测传感器。常用的位置传

are internal sensors.

2.1 Function of Internal Sensors

The main functions of the internal sensors are as follows:

(1) Detection of Specified Position and Angle

To detect a predetermined position or angle, you can use two states of on/off to detect the starting origin, limit position or exact position of the robot.

(2) Measurement of Position and Angle

Sensors that measure the linear and angular displacement of the robot joints are essential components in the robot position feedback control, including potentiometers, resolvers, and encoders.

(3) Measurement of Speed and Angular Velocity

The most common speed and angle sensors are tachogenerators or speed sensors and ratio generators. Among them, the tachogenerator that measures angular velocity can be divided into DC tachogenerator and induction AC tachogenerator according to its structure.

(4) Measurement of Acceleration

Acceleration sensors include strain gauge acceleration sensors, servo acceleration sensors, piezoelectric induction acceleration sensors, and other types of acceleration sensors.

2.2 Position Sensors

The position sensor includes position and angle detection sensors. Commonly used position sensors are analog sensors and digital sensors. The analog position sensors commonly used on industrial robots include rotary encoders, inductive synchronizer potentiometers, etc., and digital sensors include photoelectric disks, encoder disks, and gratings.

3. External Sensors

The sensors are used to detect the environment and object conditions of the robot are external sensors, such as judging the shape and spatial position of the object being grasped by the robot, whether there are obstacles around the object being grasped, and whether the object being grasped has slipped and so on.

3.1 Types of External Sensors

(1) Tactile Sensors

Generally, the sensor that detects the sense of contact, pressure, touch and proximity generated by direct contact with the outside is called a tactile sensor of robot.

(2) Force Sensors

Force perception refers to the perception of the forces experienced in the movement of the fingers, limbs and joints of the robot. It mainly includes wrist force, joint force and bearing force. According to the load of the measured object, force sensors can be divided into load sensors, torque meters, finger sensors and six-axis force sensors.

When selecting a force sensor,firstly,pay special attention to the rated value. Secondly, in the normal force control of a robot, the accuracy of the force is of little significance, and the most important thing is the resolution.

(3) Distance Sensors

Distance sensors can be used for robot navigation and obstacle avoidance, as well as for object positioning in robot space and determining its general shape characteristics. There are generally two types

感器有模拟和数字两类。常用在工业机器人上的模拟型位置传感器有旋转编码器、感应同步器、电位器等，数字型传感器有光电盘、编码盘、光栅等。

3. 外部传感器

用于检测机器人所处的环境和对象状况等的传感器是外部传感器，如判别机器人抓取对象的形状、空间位置，抓取对象周围是否存在障碍，被抓取物体是否滑落等。

3.1 外部传感器的类型

（1） 触觉传感器

一般把检测感知和外部直接接触而产生的接触觉、压力、触觉及接近触觉的传感器称为机器人触觉传感器。

（2）力觉传感器

力觉是指对机器人的指、肢和关节等运动中所受力的感知，主要包括腕力觉、关节力觉和支座力觉等。根据被测对象的负载，可以把力觉传感器分为测力传感器、力矩表、手指传感器和六轴力觉传感器。

在选用力觉传感器时，首先要特别注意额定值，其次在机器人通常的力控制中，力的精度意义不大，重要的是分辨率。

（3）距离传感器

距离传感器可用于机器人

导航和回避障碍物，也可用于机器人空间内的物体定位及确定其一般形状特征。目前常用的测距法一般有两种：激光测距法和超声波测距法。

（4）其他传感器

除以上介绍的机器人外部传感器外，还可根据机器人特殊用途安装听觉传感器、味觉传感器，而这些机器人传感器主要用于科学研究、海洋资源探测或食品分析、救火等其他特殊用途。

（5）传感器融合

为了有效地利用这些传感器信息，需要采用某种形式对传感器信息进行融合处理。不同类型信息的多种形式的处理系统就是传感器融合。传感器融合技术涉及神经网络、知识工程、模糊理论等信息、检测、控制领域的新理论和新方法。

4. 传感器的选择

传感器的选择完全取决于机器人的工作需要和应用特点，对机器人传感系统的要求是选择机器人传感器的基本依据。

（1）一般要求

机器人对传感器的一般要求有如下四点：

①精度高、重复性好。机器人能否准确无误地正常工作，往往取决于传感器的测量

of distance measurement methods commonly used at present: laser distance measurement and ultrasonic distance measurement.

(4) Other Sensors

In addition to the external sensors of the robot described above, auditory sensors and taste sensors can also be installed according to the special purpose of the robot. These robot sensors are mainly used for scientific research, marine resource detection or food analysis, fire fighting and other special purposes.

(5) Sensor Fusion

In order to effectively use these sensor information, it is necessary to use some form of synthesis and fusion processing of sensor information. Various forms of processing systems for different types of information are sensor fusion. Sensor fusion technology involves new theories and new methods in the fields of information, detection, and control such as neural networks, knowledge engineering, and fuzzy theory.

4. The Selection of Sensor

The selection of sensor depends entirely on the work needs and application characteristics of the robot. The requirements for the robot sensor system are the basis for selecting the robot sensor.

(1) General Requirements

The general requirements of robots for sensors are as follows:

①High precision and good repeatability. Whether the robot can work properly and accurately often depends on the measurement accuracy of the sensor.

②Good stability and reliability. Ensure that the

robot can work stably and reliably for a long time, and try to avoid malfunctions during work.

③Ligh anti-interference performance.The working environment of industrial robots is often harsh. The sensor used should be able to withstand certain electromagnetic interference and vibration, and be able to work normally in a high temperature, pressure and pollution environment.

④Light weight, small size and easy installation.

(2) Specific Requirements

①Adapt to the requirements of processing tasks

Different processing tasks have different requirements for robot sensors, which can be selected according to work characteristics.

②Meet the requirements of robot control

Robot control requires the use of sensors to detect the position, speed and acceleration of the robot. Most robots use position sensors as feedback elements in closed-loop control, and the robots compensate and control their motion errors according to the feedback information of the position sensors.

③Meet the requirements of the safety of the robot itself and the safety of the robot user.

精度。

②稳定性和可靠性好。保证机器人能够长期稳定可靠地工作，尽可能避免在工作中出现故障。

③抗干扰能力强。工业机器人的工作环境往往比较恶劣。其所用传感器应能承受一定的电磁干扰、振动，能在高温、高压、高污染环境中正常工作。

④质量轻、体积小、安装方便。

（2）特定要求

①适应加工任务的要求

不同的加工任务对机器人的传感器要求不同，可根据工作特点进行选择。

②满足机器人控制的要求

机器人控制需要采用传感器检测机器人的运动位置、速度和加速度。多数机器人采用位置传感器作为闭环控制中的反馈元件，机器人根据位置传感器的反馈信息对其运动误差进行补偿控制。

③满足机器人自身安全性和机器人使用者安全性的要求。

微信扫一扫

教学视频

Lecture 9-2

Perception System of Industrial Robots(2)

9-2 工业机器人的感知系统（2）

1. 位置检测传感器

1.1 绝对式角编码器

角编码器(如图 9-1 所示)是一种旋转式位置传感器，它的转轴通常与被测旋转轴连接，随被测轴一起转动（如图 9-2 所示）。角编码器能将被测轴的角位移转换成二进制编码(绝对式角编码器)或一串脉冲(增量式角编码器)。

1. Position Detection Sensor

1.1 Absolute Angle Encoder

Angle encoder (Fig. 9-1) is a kind of rotary position sensor, its rotating shaft is usually connected with the measured rotating shaft and rotates with the measured shaft, as shown in Fig. 9-2. The angle encoder can convert the angular displacement of the measured shaft into a binary code (absolute angle encoder) or a series of pulses (incremental angle encoder).

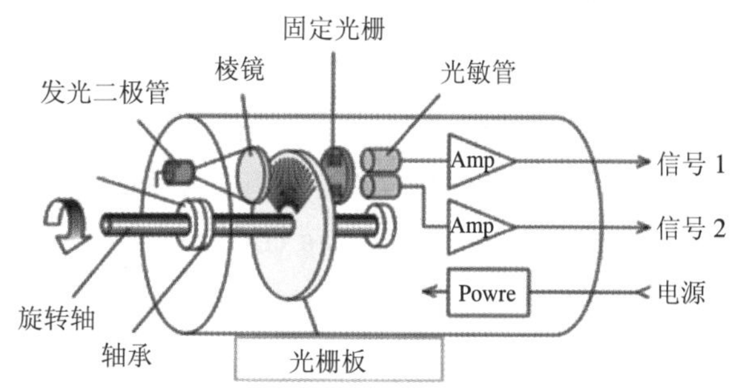

Fig.9-1 The Structure Diagram of the Angle Encoder

图 9-1 角编码器的结构示意图

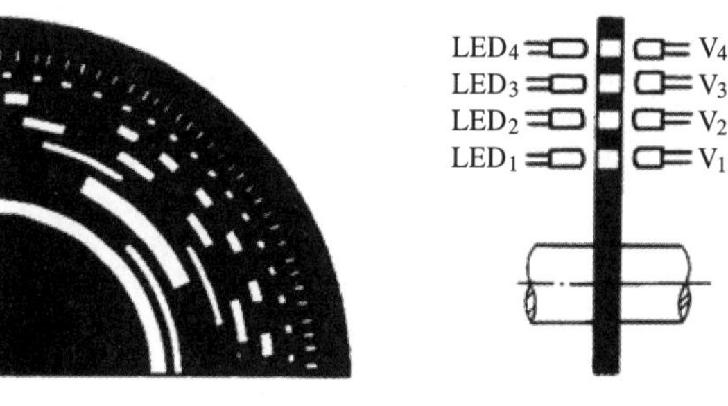

Fig.9-2 The Structure of the Absolute Angle Encoder

图 9-2 绝对式角编码器的结构

The code disc is made of optical glass,with many concentric code tracks engraved on it, thin engraved lines are deposited on the glass, and light-transmitting and opaque parts, namely bright and dark areas, are arranged in a certain pattern. When the light source projects light on the code disc, the light passing through the bright area is received by the photosensitive element. The arrangement of the photosensitive elements corresponds to the code track, the output of the photosensitive element corresponding to the bright area is "1", and the output of the dark area is "0". When the code disc is circled to different positions, the photosensitive element outputs a Gray code, which represents the angular displacement of the axis of the code disc. The stainless steel photoelectric code disc has better vibration resistance than the glass code disc. Now it also uses transparent resin film coating and etching, which has higher strength than glass code discs do.

The photoelectric code disc has no contact and wear, has a long life, and allows a high speed.

Each tiny angular displacement of the absolute angular encoder (Fig.9-3) has a corresponding code, which is often expressed in the form of binary data. In absolute measurement, even if the power is cut off in the middle, the binary coded data of the current position remains unchanged after the power is turned on again.

码盘由光学玻璃制成，其上刻有许多同心码道，在玻璃上沉积很薄的刻线，按一定规律排列的透光和不透光部分，即亮区和暗区。当光源将光投射在码盘上时，通过亮区的光线由光敏元件接收。光敏元件的排列与码道一一对应，对应于亮区的光敏元件输出为“1”，暗区的输出为“0”。当码盘旋至不同位置时，光敏元件输出格雷码，代表码盘轴的角位移的大小。

不锈钢光电码盘要比玻璃码盘抗振性好。现在也采用透明树脂片镀膜刻蚀，强度比玻璃码盘高。

光电码盘没有接触磨损，寿命长，转速高。

绝对式角编码器(如图9-3所示）的每一个微小的角位移都有一个对应的编码，常以二进制数据形式来表示。在绝对式测量中，即使中途断电，重新上电之后，当前位置的二进制编码数据仍然不变。

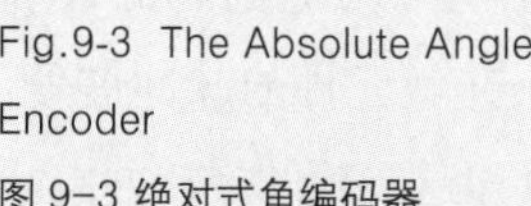

Fig.9-3 The Absolute Angle Encoder
图 9-3 绝对式角编码器

2. 速度检测传感器

2.1 增量式角编码器

增量式角编码器由转轴、发光元件、光栏板、码盘和光敏元件构成（如图 9-4、9-5 所示）。

2. Speed Detection Sensor

2.1 Incremental Angle Encoder

The incremental angle encoder is composed of a rotating shaft, a light-emitting element, an aperture plate, a code disc and a photosensitive element, as shown in Fig.9-4、9-5.

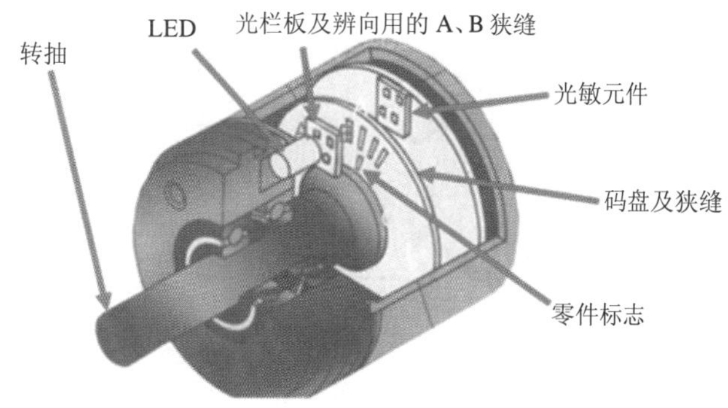

Fig. 9-4 The Structure of the Incremental Angle Encoder
图 9-4 增量式角编码器的结构

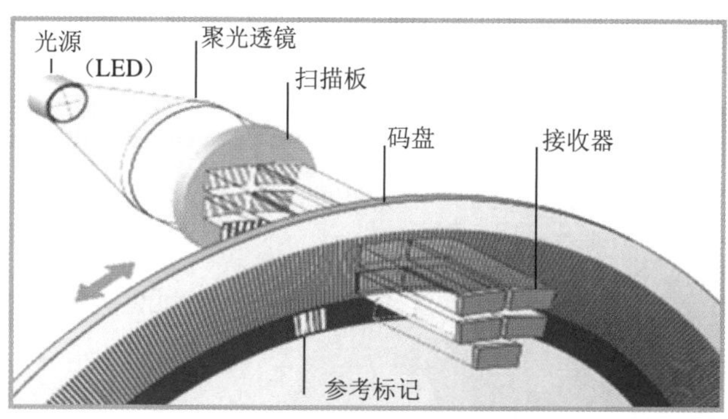

Fig. 9-5 The Structure of the Incremental Angle Encoder
图 9-5 增量式角编码器的结构

When the shaft of the incremental angle encoder rotates, the encoder will output a series of pulse signals. After counting the number of pulses, you can know how many angles the shaft has rotated in this period of time, and then know the angular velocity of the shaft. By comparing the pulse signals of the two photosensitive elements A and B, the rotation direction of the shaft can also be known.

增量式角编码器的转轴旋转时，编码器会输出一串脉冲信号，信号经过统计脉冲的个数就可以知道转轴在这段时间里转过了多少角度，进而知道转轴的角速度，通过对比 A 和 B 两个光敏元件的脉冲信号，也可以知道转轴的旋转方向。

2.2 Tachogenerator

2.2 测速发电机

The structure of a DC tachogenerator is similar to that of an ordinary generator. The stator part of the generator generates an excitation magnetic field, and the rotor is connected to the shaft. The working circuit of the DC tachogenerator is shown in the Fig. 9-6.

直流测速发电机的结构和普通发电机的结构类似，发电机的定子部分产生励磁磁场，转子和转轴连接。直流测速发电机的工作电路如图 9-6 所示。

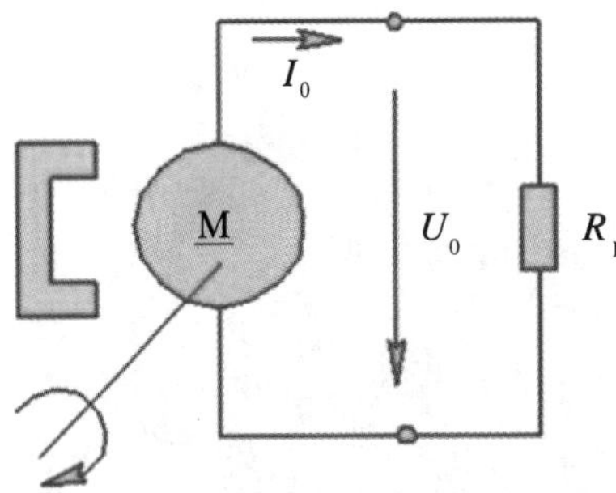

Fig.9-6 The Working Circuit of the DC Tachogenerator

图 9-6 直流测速发电机的工作电路

From the Fig. 9-7, the output voltage of the working circuit of tachogenerator is related to the linearity of the motor speed and the value of the load resistance. The larger the resistance of the load resistance is, the better the linearity of the output voltage and the motor speed is.

由图 9-7 可知，测速电机工作电路的输出电压和电机转速的线性度和负载电阻的大小有关，负载电阻的阻值越大，输出电压和电机转速的线性度越好。

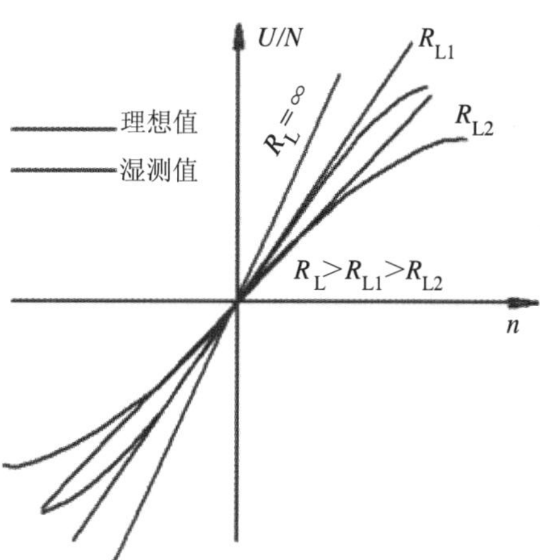

Fig. 9-7 The Relationship Between the Output Voltage and the Motor Speed of the Tachogenerator Circuit

图 9-7 测速电路的输出电压和转速关系

3. 加速度传感器

3.1 应变式加速度传感器

通过测量惯性力引起弹性敏感元件的变形量换算成力的关系，间接度量加速度。应变式加速度传感器的结构如图9-8所示。

3. Acceleration sensor

3.1 Strain Accelerometer

The acceleration is indirectly measured by measuring the deformation of the elastic sensitive element caused by the inertial force and converting it into force. The structure of the strain accelerometer is shown in the Fig.9-8.

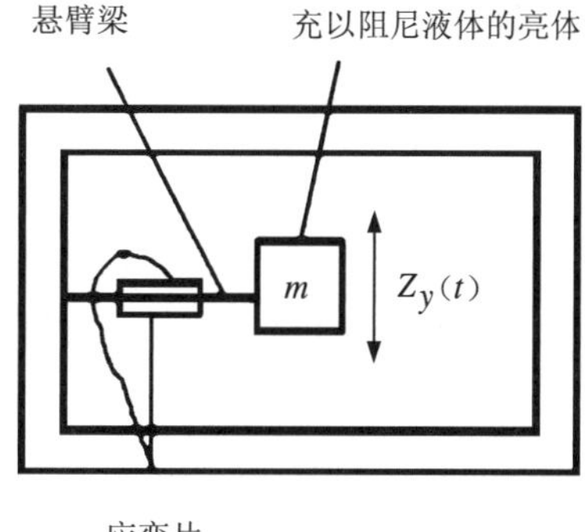

Fig. 9-8 The structure of the strain accelerometer

图 9-8 应变式加速度传感器的结构

Lecture 10-1
Control System of Industrial Robots(1)

微信扫一扫

教学视频

10-1 工业机器人的控制系统（1）

Industrial robot controllers have powerful processing functions and rich input/output interfaces, which can be connected to motion control boards, and can be networked with other controllers to form an intelligent control network.

1. The Function of Industrial Robot Control System

The main task of the industrial robot control system is to control the movement position, posture and trajectory, operation sequence and action time of the industrial robot in the working space. The control of some projects is very complicated, which determines that the control system of industrial robots should have the following characteristics:

(1) The control system of industrial robots must be used to often solve kinematics and dynamics problems.

(2) The control system of industrial robots is not only a nonlinear system, but also a multivariable system.

(3) The control system of industrial robots must be used to also solve optimization problems.

2. General Requirements for Control System

(1) Memory function: storage of work sequence,

工业机器人控制器具有强大的处理功能以及丰富的输入/输出接口，能够连接运动控制板卡，能够和其他的控制器进行组网，构成一个智能控制网络。

1. 工业机器人控制系统的功能

工业机器人控制系统的主要任务是控制工业机器人在工作空间中的运动位置、姿态和轨迹、操作顺序及动作的时间等。有些项目的控制是非常复杂的，这就决定了工业机器人的控制系统应具有如下特点：

（1）工业机器人的控制系统要经常求解运动学和动力学问题。

（2）工业机器人的控制系统不仅是一个非线性系统，而且是一个多变量系统。

（3）工业机器人的控制系统还必须解决优化问题。

2.对控制系统的一般要求

（1）记忆功能：存储作业顺序、运动路径、运动方式、运动速度和与生产工艺有关的信息。

（2）示教功能：离线编程、在线示教、间接示教。

（3）与外围设备联系功能：输入和输出接口、通信接口、网络接口、同步接口。

（4）坐标设置功能：有关节、绝对、工具、用户自定义4种坐标系。

（5）人机接口：示教盒、操作面板、显示屏。

（6）传感器接口：位置检测、视觉、触觉、力觉等。

（7）位置伺服功能：机器人多轴联动、运动控制、速度和加速度控制、动态补偿等。

（8）故障诊断安全保护功能：运行时系统状态监视、故障状态下的安全保护和故障自诊断。

3. 机器人控制系统类型

机器人控制系统的类型分为程序控制系统、自适应控制系统和人工智能系统。

程序控制系统：给每一个自由度施加一定规律的控制作用，机器人就可实现所要求的空间轨迹。

自适应控制系统：当外界

motion path, motion mode, motion speed and information related to production technology.

(2) Teaching function: offline programming, online teaching and indirect teaching.

(3) Contact with peripheral equipment: input and output interface, communication interface, network interface and synchronization interface.

(4) Coordinate setting function: There are 4 coordinate systems of joint, absolute, tool, and user-defined.

(5) Man-machine interface: teaching box, operation panel and display screen.

(6) Sensor interface: position detection, vision, touch, force, etc.

(7) Position servo function: robot multi-axis linkage, motion control, speed and acceleration control, dynamic compensation, etc.

(8) Fault diagnosis safety protection function: system status monitoring during operation, safety protection under fault status and fault self-diagnosis.

3.The Type of Robot Control System

The types of robot control systems are divided into:program control systems,adaptive control systems and artificial intelligence systems.

Program control system: By applying a certain regular control function to each degree of freedom, the robot can realize the required space trajectory.

Adaptive control system: When the external conditions change, in order to ensure the required quality or to improve the control quality with the accumulation of experience, the process is based on the observation of the state of the operating machine and the servo error, and then adjust the nonlinear model parameters until the error disappears.

Artificial intelligence system: It is impossible to compile exercise programs in advance, but requires real-time determination of the control functions according to the obtained surrounding state information during exercise.

4. The Structure of the Robot Control System

(1) Centralized Control Method

The centralized control method uses one computer to realize all control functions, with simple structure and low cost, but the real-time performance is poor and is difficult to expand, as shown in Fig. 10-1.

条件变化时，为保证所要求的品质或为了随着经验的积累而自行改善控制品质，其过程是基于操作机的状态和伺服误差的观察，再调整非线性模型的参数，一直到误差消失为止。

人工智能系统：事先无法编制运动程序，而是要求在运动过程中根据所获得的周围状态信息实时确定控制作用。

4. 机器人控制系统结构

（1） 集中控制方式

集中控制方式用一台计算机实现全部控制功能，结构简单，成本低，但实时性差，难以扩展，如图 10-1 所示。

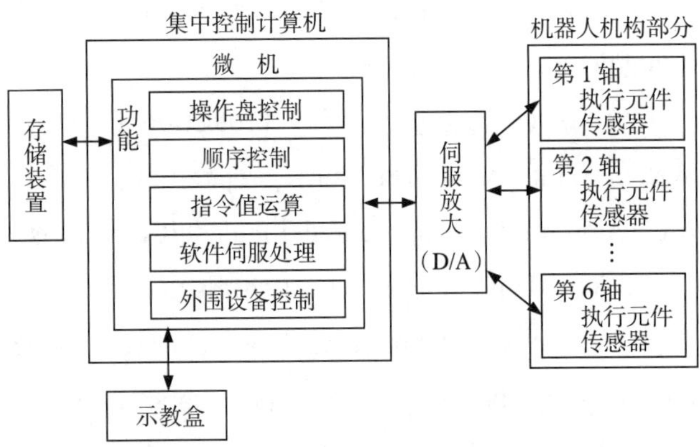

Fig.10-1 Centralized Control Method

图 10-1 集中控制方式

(2) Master-Slave Control Method

The master-slave control method adopts master and slave two-level processors to realize all control functions of the system. The main CPU realizes management, coordinates transformation, trajectory generation, and system self-diagnosis, and the slave CPU realizes the motion control of all joints, as shown in Fig.10-2.

（2） 主从控制方式

主从控制方式采用主、从两级处理器实现系统的全部控制功能。主 CPU 实现管理、坐标变换、轨迹生成和系统自诊断等，从 CPU 实现所有关节的动作控制，如图 10-2 所示。

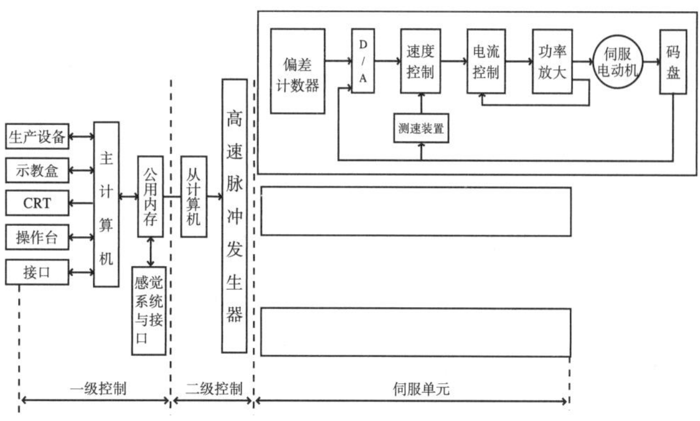

Fig.10-2 Master-Slave Control Method
图 10-2 主从控制方式

(3) 分散控制方式

分散控制方式按系统的性质和方式将系统控制分成几个模块，它们有不同的控制任务和控制策略，各模块之间可以是主从关系，也可以是平等关系，如图 10-3 所示。

(3) Decentralized Control Method

The decentralized control method divides the system control into several modules according to the nature and method of the system. They have different control tasks and strategies, and each module can have a master-slave relationship or an equal relationship, as shown in Fig. 10-3.

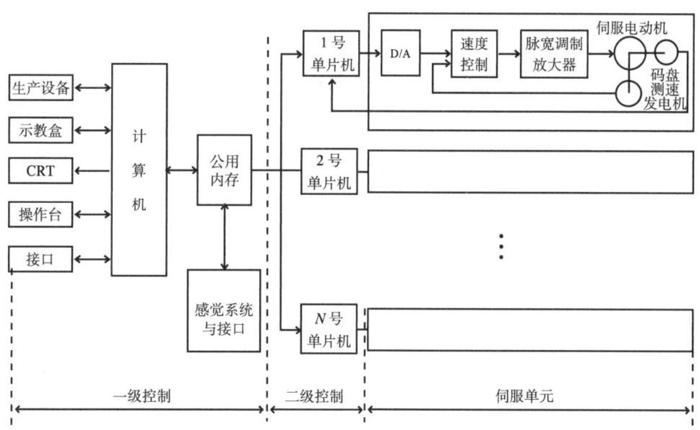

Fig.10-3 Decentralized Control Method
图 10-3 分散控制方式

①工业机器人的控制与其机构运动学和动力学有着密不可分的关系，所以要使工业机器人的臂、腕及末端执行器等部位在空间具有准确无误的位姿，就必须在不同的坐标系中定义它们，而且随着基准坐标系的不同要做适当的坐标变换。

① The control of industrial robots is inseparable from the kinematics and dynamics of its mechanism. Therefore, in order to make the arms, wrists and end effectors of industrial robots have accurate poses in space, they must be described in different coordinate systems, and appropriate coordinate transformations must be made according to the different reference coordinate systems.

② The mathematical model describing the state and movement of the industrial robot is a nonlinear model, which changes with the movement and environment of the industrial robot. And because industrial robots often have multiple degrees of freedom, there is more than one variable that causes its motion to change, and there are coupling problems among each variable.

③ Any pose of the industrial robot can be reached through different ways and paths, the master-slave control method has good real-time performance and is suitable for high-precision and high-speed control, but its system scalability is poor and maintenance is difficult.

This method has good real-time performance. It is easy to realize high-speed, high-precision control, easy to expand, and can realize intelligent control. It is a popular method at present.

②描述工业机器人状态和运动的数学模型是一个非线性模型，随着工业机器人的运动及环境改变。又因为工业机器人往往具有多个自由度，所以引起其运动变化的变量不止一个，而且各个变量之间存在耦合问题。

③工业机器人的任何位姿都可以通过不同的方式和路径到达，主从控制方式系统实时性较好，适用于高精度、高速度控制，但其系统扩展性较差，维修困难。

这种方式实时性较好，易于实现高速、高精度控制、易于扩展，可实现智能控制，是目前流行的方式。

Lecture 10-2

Control System of Industrial Robots(2)

微信扫一扫

教学视频

10-2 工业机器人的控制系统（2）

1. 控制系统的硬件

由于安全及运行过程的需要，工业机器人的控制系统必须具有实时控制功能。因此，工业机器人控制系统除了有明确的软、硬件分工外，更重要的是应具有实时操作系统。另外，工业机器人的许多任务既可用硬件完成，也可用软件完成。这些任务的实现手段（采用硬件还是软件完成）主要取决于执行速度、精度要求及实现方式的难易程度（结构、成本及维护等）。

一般工业机器人控制系统的软、硬件任务分配如下：速度平滑控制、自动加减速控制及防振控制采用专用软件方式处理；硬件系统应配合其他软件完成以下模块功能：

①系统控制。

②示教操作、编程与 CRT 显示。

③多轴控制、速度协调控

1. Hardwares of the Control System

Due to the requirements of safety and operation process, the control system of industrial robots must have real-time control functions.Therefore,in addition to a clear division of software and hardware, the industrial robot control system should also have a real-time operating system. Besides, many tasks of industrial robots can be completed with hardware or software. The means of achieving these tasks (using hardware or software) mainly depends on the execution speed, accuracy requirements, and the difficulty of the implementation (structure, cost and maintenance, etc.).

Generally, the software and hardware tasks of the industrial robot control system are distributed as follows: speed smoothing control, automatic acceleration and deceleration control and anti-vibration control are processed by special software; the hardware system should cooperate with other software to complete the following module functions:

① System control.

② Teaching operation, programming and CRT display.

③ Multi-axis control, speed coordinated control (reproduction).

④ I/O communication and interface control.

⑤ Various security and interlocking control.

Hardware Composition of the Control System

The controllers of the robot control system mostly use industrial control computers, PLCs, single-chip microcomputers or single-board computers, etc. In recent years, they are gradually developing towards open numerical control systems.

The advantage of interface mode of general functions is that it can flexibly deal with different numbers of sensors and realize the generalization of various circuit boards.

Interface serialization can simplify the connection between devices, and standardize the physical conditions or protocols of the interface, and the utilization value of the interface will be greatly improved. At present, basically only the LAN (Local Area Network) method is adopted. The information exchange between LANs with the help of memory can easily realize the connection between different LANs (so-called smart connectors). Here the calculation result of the controller is placed in the dual-channel memory, and the robot reads the program execution result in the dual-channel memory through the network connector.

A robot using a PC controller full set of axes use an integrated servo driver. The driver interface is formed by ARCnet, and a dedicated motion control board can be attached for high-speed calculation.

制（再现）。

④I/O 通信与接口控制。

⑤各种安全与连锁控制。

控制系统的硬件组成

机器人控制系统的控制器多采用工业控制计算机、PLC、单片机或单板机等，近年来正逐渐向开放数控系统发展。

通用功能的接口方式的优点是可以灵活地应对不同数量的传感器，实现各种电路板卡的通用化。

接口串行化能简化设备之间的连接，将接口的物理条件或协议标准化，那么接口的利用价值就会大幅度提高。目前，基本上仅采用 LAN（局域网）方式，借助于存储器进行 LAN 之间的信息交换，能方便地实现不同方式的 LAN 之间的连接（所谓的智能连接器）。这里控制器的运算结果放在双通道存储器中，机器人通过网络连接器读取双通道存储器里的程序执行结果。

采用 PC 控制器的机器人，全套轴采用一个集成的伺服驱动器，由 ARCnet 构成驱动器接口，可以附加专用运动控制板卡进行高速计算。

2. 控制系统的软件

2.1 控制系统软件的功能

工业机器人的柔性体现在其运动轨迹、作业条件和作业顺序能自由变更，变更的灵活程度取决于其软件的功能水平。

工业机器人按照操作人员的示教动作及要求进行作业，操作人员可以根据作业结果或条件进行修正，直到满足要求为止。因此，软件系统应具有以下基本功能：

①示教信息输入；

②对机器人及外部设备动作的控制；

③运行轨迹在线修正；

④实时安全检测。

2.2 控制系统软件的构成

控制系统软件主要由三部分构成：

①操作系统

操作系统是硬件系统相关的程序集合，用于协调控制器内部任务，也提供同外部通信的媒介。其任务同计算机操作系统相类似，包括主存储器处理、接收和发送数据、输入输出单元、外围设备、传感器输入设备及对其他通信要求的响应；对于规模较小的控制系统则采用监控系统。

2. Software of the Control System

2.1 Functions of the Control System Software

The flexibility of an industrial robot is reflected in its movement trajectory, operating conditions and operating sequence can be freely changed, and the flexibility of the change depends on the functional level of its software.

Industrial robots operate in accordance with the teaching actions and requirements of the operator, and the operator can make corrections according to the results or conditions of the operation until the requirements are met. Therefore, the software system should have the following basic functions:

① Input of teaching information;

② Control of the actions of robots and external equipment;

③ Online correction of running track;

④ Real-time safety inspection.

2.2 Composition of the Control System Software

The control system software is mainly composed of three parts:

① Operating system

The operating system is a collection of programs related to the hardware system, used to coordinate the internal tasks of the controller, and also provide a medium for communication with the outside. Its tasks are similar to computer operating systems, including main memory processing, receiving and sending data, input and output units, peripherals, sensor input devices and response to other communication requirements; for smaller control systems, monitoring systems are used.

② Special program modules for robots

The special program modules for robots include coordinate transformation, which transmits application special commands to the manipulator, provides trajectory generation, kinematics and dynamics constraints, handles force feedback, speed control, vision input and other sensor inputs, and processes input data and machine-level I/O errors.

③ Robot language

Robot language is a software interface through which programmers can directly manipulate the robot to perform required actions. This language should have a user-friendly interface, provide simple editing functions, and use macro instructions or subroutines to solve application specific tasks.

②机器人专用程序模块

机器人专用程序模块包括坐标变换，为操作机传递应用的特殊命令，提供轨迹生成、运动学和动力学的限制条件，处理力反馈、速度控制、视觉输入和其他传感器输入，处理输入数据和面向机器级的 I/O 错误。

③机器人语言

机器人语言是软件接口、编程者通过它可直接操纵机器人执行需要的动作。这种语言应具有友好的用户界面，提供简单的编辑功能，可使用宏指令或子程序解决应用的具体任务。

Lecture 11-1
Programming and Debugging of Industrial Robot System(1)

微信扫一扫

教学视频

11-1 工业机器人的编程和调试（1）

ABB 机器人的 RAPID 编程提供了丰富的指令来完成各种简单与复杂的应用。

任务要求：通过工业机器人 tool1 工具的 TCP 点在白板上绘制一个五角星，要求从 Home 机械原点出发，绘制完成后再回到 home 机械原点，并且工具的出入点要有偏移量。

1. 创建例行程序

首先，单击示教器左上角主菜单按钮→选择“程序编辑器”→单击“取消”→单击左下角“文件”菜单→单击文件中的“新建模块”→单击“是”→设定模块名称（这里就使用默认名称 Module1），点击“确定”→选中 Module1，单击“显示模块”→单击“例行程序”→单击左下角“文件”菜单→单击文件里的“新建例行程序”→设定例行程序名称（这里就使用默认名称 Routine1），单击“确定”→选中 Routine1，单击

The RAPID programming of ABB robots provides a wealth of instructions to complete various simple and complex applications.

Task requirements: draw a five-pointed star on the white board by robot through the TCP point of the tool1 by the industrial robot. It is required to start from mechanical origin and return to the mechanical origin after the drawing is completed. The entry and exit points of the tool must be offset.

1. Create Routine

First, click the main menu button in the upper left corner of the Flexpendant → select “Program Editor” → click “Cancel” → click the “File” menu in the lower left corner → click “New Module” in the file → click “Yes” →Set the module name (here, use the default name Module1), and click “OK” → select Module1, click “Display Module” → click “Routine” → click the “File” menu in the lower left corner → click the file “New Routine” → set the name of the routine (here the default name Routine1 is used), and click“OK”→ select Routine1,click “Display Routine” → select the position of the program where you want to insert the instruction, highlight it in blue → click

"Add Instruction" to open the instruction list → click the common button to switch to the instruction list of other categories.

2. Add Mechanical Origin Instructions

Select "<SMT>" as the location to add instructions → select "MoveAbsJ" in the instruction list → select the "*" and highlight it in blue, then click the "*" → click "New" → click "..." to change the name to home and click "OK" → click "OK" after rewriting the name → "*" has been replaced by the variable of the home target point, select the home point and click "OK".

3. Add Linear Motion Instructions

Select "MoveL" in the instruction list → click "Below" → select the "*" and highlight it in blue, and then click the "*" → click "New" → change the name to the default p10, and click "OK" → Select p10, click "OK" → use the same method to create p20, p30, p40, p50 in sequence.

4. Set the Offset

Select the "MoveAbsJ home" instruction program → click "Edit" → click "Copy" → select the MoveL p50 instruction program → click "Paste" → select the MoveL p10 instruction program, and use the same method to copy a line below it. Select the first "p10" and highlight it in blue, then click "p10" → select "p10", then click "Functions" → click "Offs" → click "p10" → select the first < EXP>, and click "Edit" → click "Selected content only" → enter "0", and click "OK" → following the same method, change the next two <EXP> to 0 and 100 → click "OK" → click "OK" again → use the same method to copy a

"显示例行程序"→选中要插入指令的程序位置，高显为蓝色→单击"添加指令"打开指令列表→单击 common 按钮可切换到其他分类的指令列表。

2. 添加机械原点指令

选中"<SMT>"为添加指令的位置→在指令列表中选择"MoveAbsJ"→选中*号并蓝色高亮显示，再单击*号→单击"新建"→单击"…"将名称改为 home 后单击"确定"→名称改写完成后单击"确定"→*号已经被 home 目标点变量代替，选中 home 点单击"确定"。

3. 添加直线运动指令

在指令表中选择"MoveL"→单击"下方"→选中*号并蓝色高亮显示，再单击*号→单击"新建"→将名称改为默认 p10，单击"确定"→选中 p10，单击"确定"→用同样的方法依次建立 p20、p30、p40、p50。

4. 设置偏移量

选中"MoveAbsJ home"指令程序→单击"编辑"→单击"复制"→选中 MoveL p50 指令程序→单击"粘贴"→选中 MoveL p10 指令程序，用同样的方法在其下方复制一行。选中第一个"p10"并蓝色高亮显示，再单击"p10"→选中

"p10"，再单击"功能"→单击"Offs"→单击"p10"→选中第一个<EXP>，单击"编辑"→单击"仅限选定内容"→输入"0"，单击"确定"→依照同样的方法，将后面两个<EXP>更改为 0 和 100→单击"确定"→再次单击"确定"→用同样的方法将MoveL p50指令程序在下方复制一行→用同样的方法将复制的 MoveL p50指令程序中的 P50 设定增量为 100。

5. 示教位置点

选中第一行的 home 点→单击"调试"→单击"查看值"→将 rax_1～rax_6 的值设置成 0、0、0、0、90、0→单击"确定"→将机器人的工具原点移动到五角星的第一个点，选中"P10"然后单击"修改位置"→用同样的方法将机器人工具原点分别移动到五角星的第二、第三、第四、第五个点，分别选中 p20、p30、p40、p50，并修改位置。

6. 调试

单击"调试"→单击"PP移至例行程序"→选中"Routinel1"→单击"确定"→左手四指按下使能按键，确定那个电机开启，单步运行程序，确保无误后，再连续运行程序。

line of MoveL p50 instruction program below → use the same method to set the increment of P50 in the copied MoveL p50 instruction program to 100.

5. Teaching Position Point

Select the home point in the first row→ click "Debug"→ click "View Value"→ set the value of rax_1～rax_6 to 0, 0, 0, 0, 90, 0 → click"OK"→ move the tool origin of the robot to the first point of the five-pointed star, select "P10" and click "Modify Position"→ use the same method to move the robot tool origin to the second, third, fourth, and fifth point of the five-pointed star, respectively select p20, p30, p40, p50, and modify the position.

6. Debug

Click "Debug"→ click "Move PP to Routine" → select "Routinel1"→ click "OK"→ Press the enabler button with four fingers of your left hand to confirm that the motor is turned on, and run the program in a single step, and then run the program continuously after ensuring that it is correct.

微信扫一扫

教学视频

Lecture 11-2
Programming and Debugging of Industrial Robot System(2)

11-2 工业机器人的编程和调试（2）

After editing the program, it is usually necessary to debug the program. There are two purposes of debugging: one is to check whether the point in the program is correct; the second is to check whether the logic control in the program is reasonable and complete.

在完成程序的编辑后，通常需要对程序进行调试，调试的目的有两个：一是检查程序中位置点是否正确；二是检查程序中的逻辑控制是否合理和完善。

1.Debug the rHome Routine

1. 调试 rHome 例行程序

As shown in Fig.11-1

如图 11-1 所示。

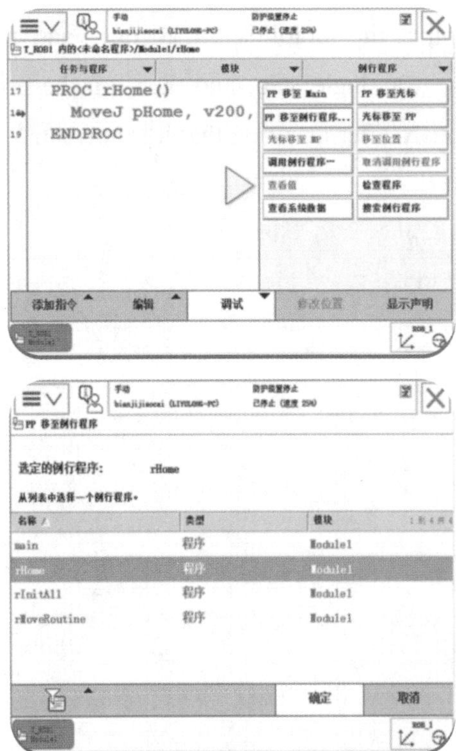

Fig.11-1 Debug the rHome Routine

图 11-1 调试 rHome 例程

①打开调试菜单，单击“PP 移至例行程序”。

②选中 rHome 例行程序，然后单击“确定”。

①Open the debug menu and click “Move PP to Routine…”.

②Select the rHome routine and click “OK”.

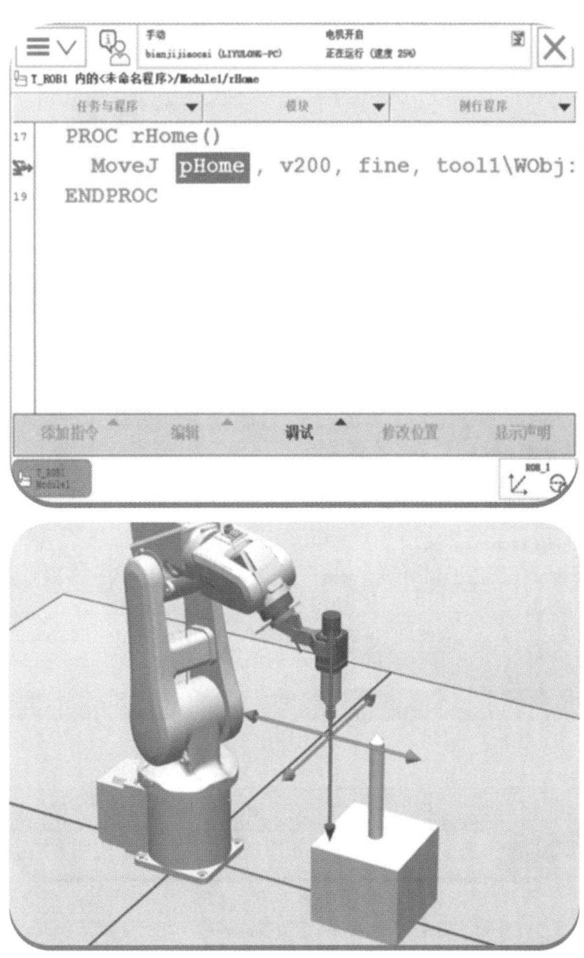

Fig.11-2 Move Routine

图 11-2 移动例程

③按下示教器使能键，按一下单步向前按键，当程序指针（黄色小箭头）与小机器人图标指向同一行时，说明机器人已到达 pHome 点位置。

④此时观察真实环境中机器人的位置是否与用户定义的 pHome 点位置一样。

③Press the enabler button of the Flexpendant, and press the single step forward button. When the program pointer (small yellow arrow) and the small robot icon point to the same line, it indicates that the robot has reached the position of pHome point.

④At this time, observe whether the position of the robot in the real environment is the same as the position of the user-defined pHome point.

2. Debug the rMoveRoutine Routine

2. 调试 rMoveRoutine 例行程序

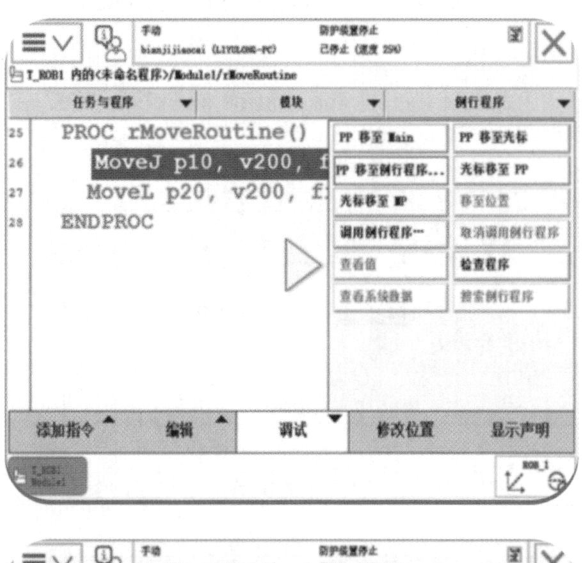

Fig.11-3 Debug the rMoveRoutine Routine 1

图 11-3 调试 rMoveRoutine 例程 1

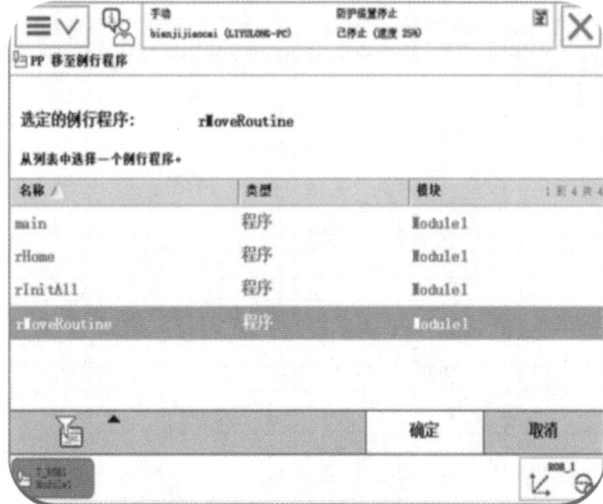

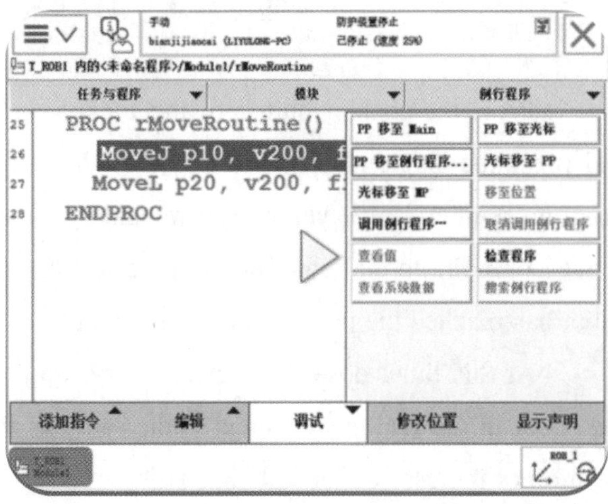

Fig.11-4 Debug the rMoveRoutine Routine2

图 11-4 调试 rMoveRoutine 例程 2

①打开调试菜单，单击“PP 移至例行程序”。

②选中 rMoveRoutine 例行程序，然后单击“确定”。

①Open the debug menu and click “Move PP to Routine…”.

②Select rMoveRoutine routine, and then click “OK”.

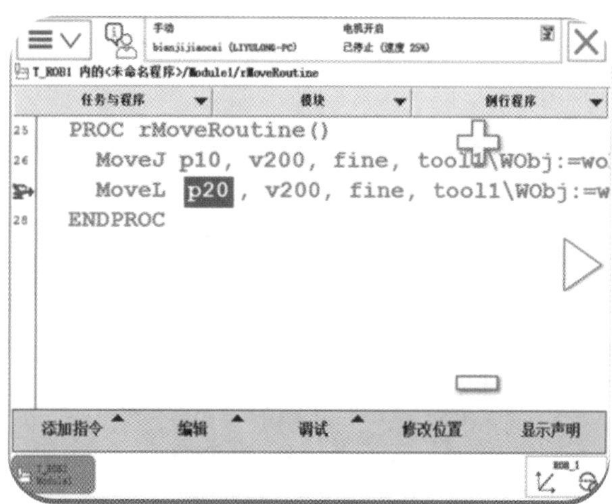

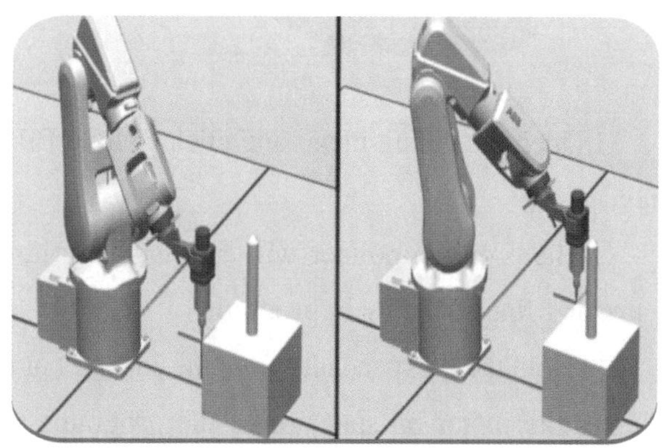

Fig.11-5 Move Routine

图 11-5 移动例程

③按下示教器使能键，按一下单步向前按键，当程序指针（黄色小箭头）与小机器人图标指向同一行时，说明机器人已到达程序中的位置。

④在进行单步调试过程中，可观察每一点的位置是否合适。

③Press the enabler button of the Flexpendant, and press the single step forward button. When the program pointer (small yellow arrow) and the small robot icon point to the same line, it indicates that the robot has reached the position in the program.

④In the process of single-step debugging, you can observe whether the position of each point is appropriate.

3. Debug the"main"Main Program

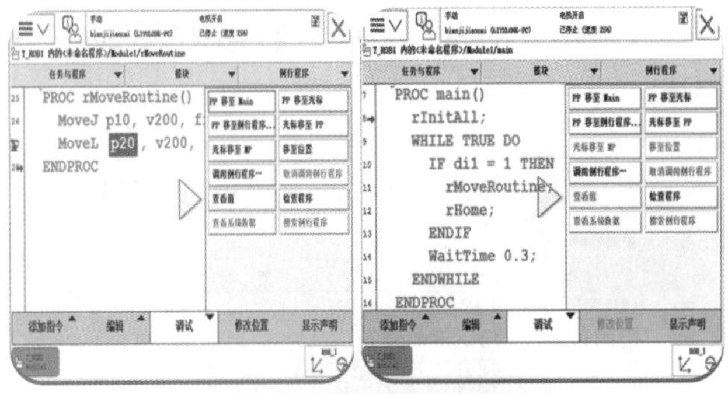

Fig.11-6 Debug the"main"Main Program

图 11-6 调试 main 主程序

①Open the debug menu and click "Move PP to Main".

②The program pointer will automatically jump to the first line of the main program.

③ Hold the Flexpendant, press the enabler button to enter the motor-on state, press the program start button, and carefully observe the movement of the robot. If you need to stop the robot during the process, you must first press the program stop button, and then release it.

4. Automatic Operation of the Program

In the manual state, after completing the debugging of the robot program, the robot can be put into the automatic operation state. Its automatic operation is as follows:

3. 调试 main 主程序

①打开调试菜单，单击"PP 移至 Main"。

②程序指针会自动跳至主程序的第一行指令。

③手持示教器，按下使能键，进入电机开启状态，按一下程序启动按键，并小心观察机器人的移动，若过程中需要停止机器人，务必先按下程序停止按键，然后再松开使能键。

4. 程序的自动运行

在手动状态下，完成了对机器人程序的调试后，就可以将机器人投入自动运行状态。其自动运行的操作如下：

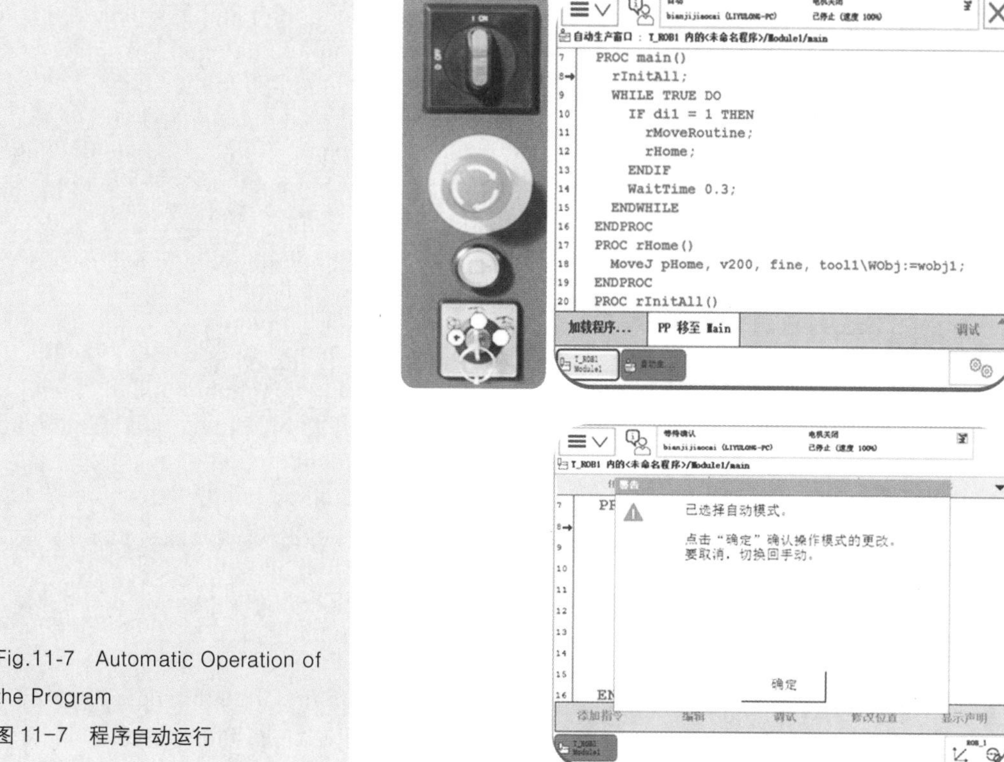

Fig.11-7 Automatic Operation of the Program

图 11-7 程序自动运行

①将状态钥匙逆时针旋转至自动状态。

②在示教器界面上，单击“确定”，确认状态的切换。

③单击 PP 移至 Main，将程序指针指向主程序的第一行指令。

①Turn the status key counterclockwise to the automatic status.

②On the Flexpendant interface, click “OK” to confirm the status switch.

③ Click “Move PP to Main”, and point the program pointer to the first line of the main program.

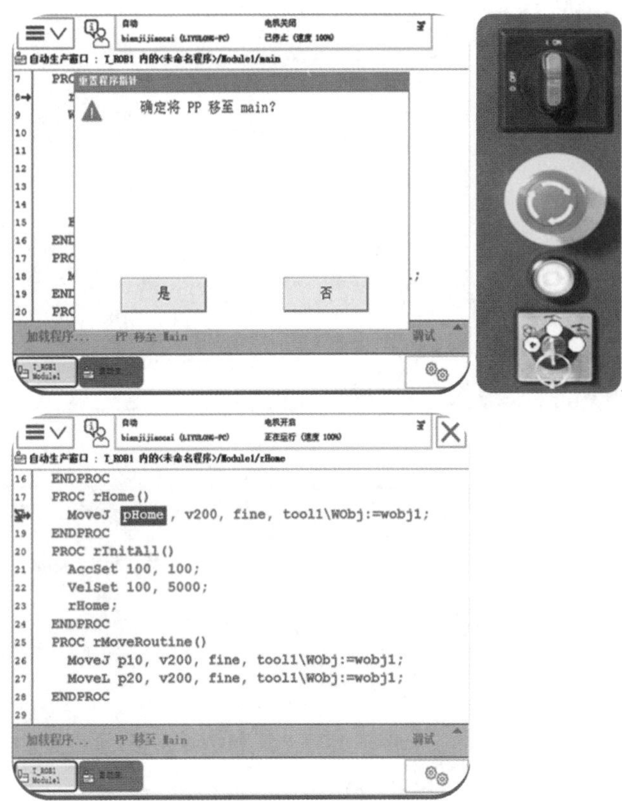

Fig.11-7 Switch operation state

图 11-7 切换运行状态

④Click “Yes”.

⑤Press the white button to turn on the motor.

⑥Press the program start button to observe the robot program start to run automatically.

④单击“是”。

⑤按下白色按钮，使电机处于开启状态。

⑥按下程序启动按键，可以观察机器人程序开始自动运行。

微信扫一扫

教学视频

Lecture 12-1
Knowledge of Industrial Robot Workstations (1)

12-1 了解机器人工作站（1）

现代制造业正朝着无人化、智能化、个性化和绿色化的方向发展，传统机械加工设备的智能化程度不足以满足未来发展需求。由工业机器人、PLC控制系统和各种智能化设备通过总线组网构成的工作站或生产线正成为制造业的主流设备。

1. 工业机器人工作站的概念

机器人工作站有别于传统的自动化生产线，是指以一台或多台工业机器人为主，具有强大的控制系统，友好的人机交互界面，配以相应的末端执行器及周边的设备（如变位机、输送机、工装夹具等），有些情况下还可以借助人工辅助操作一起完成相对独立的一种作业或工序的柔性工作站。如图12-1所示是汽车焊接工作站。图12-2是零件摆放工作站。图12-3是水果装箱工作站。图12-4是包裹转运工作站。图

The modern manufacturing industry is developing towards unmanned,intelligent, personalized and green direction. The intellectualization degree of traditional mechanical processing equipment is not enough to meet the future development needs. Workstations or production lines composed of industrial robots, PLC control systems and various intelligent equipment through bus networking are becoming mainstream equipment in the manufacturing industry.

1. The concept of industrial robot workstation

Robot workstations are different from traditional automated production lines. They are based on one or more industrial robots, with powerful control systems, friendly human-machine interaction interfaces, and corresponding end effectors and peripheral equipment (such as deflection machine, conveyor, tooling fixture, etc.), in some cases, it is also possible to use the flexible workstation to complete a relatively independent work or process with the help of manual assisted operation. The Automobile Welding Workstation is shown in Fig.12-1. Fig.12-2 is a Parts Placement Workstation. Fig.12-3 is a Fruit Packing Workstation. Fig.12-4 is a Package Transfer Workstation. Fig. 12-5 is a Part

12-5 是零件冲压工作站。

Fig.12-1 The Automobile Welding Workstation
图 12-1 汽车焊接工作站

Fig.12-2 Parts Placement Workstation
图 12-2 零件摆放工作站

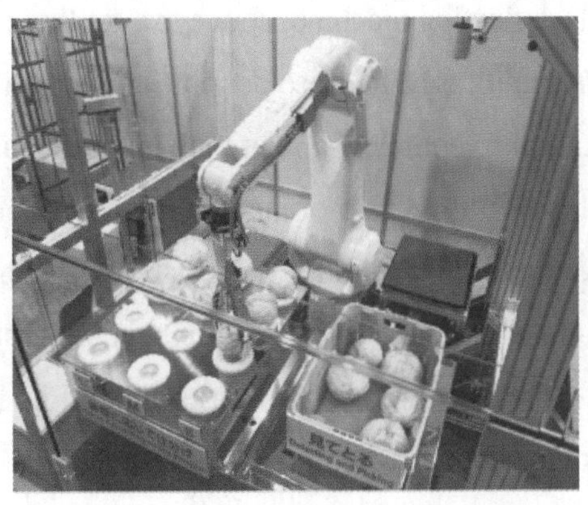

Fig.12-3 Fruit Packing Workstation
图 12-3 水果装箱工作站

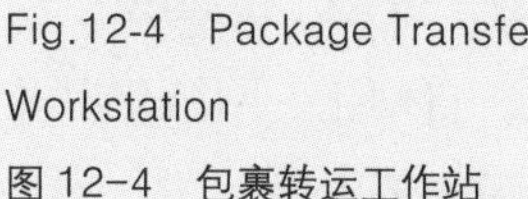

Fig.12-4 Package Transfer Workstation
图 12-4 包裹转运工作站

Fig.12-5 Part Stamping Workstation
图 12-5 零件冲压工作站

Stamping Workstation.

2. 机器人工作站的特点

（1）工业机器人工作站技术先进

工业机器人集精密化、柔性化、智能化、软件应用开发等先进制造技术于一体，通过对过程实施检测、控制、优化、调度、管理和决策，实现增加产量、提高质量、降低成本，减少资源消耗和环境污染的目的，是工业自动化高水平的体现。

（2）工业机器人工作站技术升级

2.The characteristics of the robot workstation

(1)The industrial robot workstation has advanced technology

Industrial robots integrate advanced manufacturing technologies such as precision, flexibility, intelligence, and software application development. Through the implementation of process detection, control, optimization, scheduling, management and decision-making, to achieve the purpose of increasing output, improving quality, reducing costs, reducing resource consumption and environmental pollution, is the embodiment of the high level of industrial automation.

(2) The industrial technology of the robot workstation has upgraded

Industrial robots and automated complete sets of equipment have technical characteristics such as fine manufacturing, fine processing, and flexible production. They are new-generation production tools that fully extend human physical and intellectual capabilities after power machinery and computers, and they are important means to realize the digitalization, automation, networking and intelligentization of production.

(3) The application of industrial robot workstations is widespread

Industrial robots and complete sets of automation equipment are the key equipment in the production process, which can be used in manufacturing, installation, testing, logistics and other production links. They are widely used in automobiles and auto parts, construction machinery, rail transit, low-voltage electrical appliances, electricity, IC equipment, military industry, tobacco, finance, medicine, metallurgy, printing and publishing industries have a wide range of applications;

(4)The technology of industrial robot workstation is comprehensive

Industrial robots and complete sets of automation technologies concentrate and integrate a number of disciplines, involving a number of technological areas,including advanced manufacturing technologies such as industrial robot control technology, robot dynamics and simulation, robot construction finite element analysis, laser processing technology, modular programming, intelligent measurement, integration of modeling and processing, factory automation and fine logistics, which are comprehensive.

工业机器人与自动化成套装备具有精细制造、精细加工以及柔性生产等技术特点，是继动力机械、计算机之后出现的全面延伸人的体力和智力的新一代生产工具，是实现生产数字化、自动化、网络化以及智能化的重要手段。

（3）工业机器人工作站应用领域广泛

工业机器人与自动化成套装备是生产过程的关键设备，可用于制造、安装、检测、物流等生产环节，并广泛应用于汽车整车及汽车零部件、工程机械、轨道交通、低压电器、电力、IC 装备、军工、烟草、金融、医药、冶金及印刷出版等行业，应用领域非常广泛。

（4）工业机器人工作站技术综合性强

工业机器人与自动化成套技术集中并融合了多项学科，涉及多项技术领域，包括工业机器人控制技术、机器人动力学及仿真、机器人构建有限元分析、激光加工技术、模块化程序设计、智能测量、建模加工一体化、工厂自动化以及精细物流等先进制造技术，技术综合性强。

3. 工业机器人工作站的结构

工业机器人工作站是以工业机器人作为加工主体的作业系统。由于工业机器人具有可再编程的特点，当加工产品更换时，可以对机器人的作业程序进行重新编写，从而达到系统柔性要求。

然而，工业机器人只是整个作业系统的一部分，作业系统还包括工装、变位器、辅助设备等周边设备，工业机器人工作站对工业机器人和周边设备进行系统集成，在硬件上根据需求对各个设备接口进行统一定义，以满足通信需求，在软件上对整个系统的信息流进行综合，然后再控制各个设备按流程运转。

点焊机器人工作站如图12-6所示。

3. The structure of the industrial robot workstation.

The industrial robot workstation is an operating system with industrial robots as the main processing body. Due to the reprogrammability of industrial robots, when the processed products are changed, the robot's operating procedures can be rewritten to meet the requirements of system flexibility.

However, industrial robots are only part of the entire operating system. The operating system also includes peripheral equipment such as tooling, positioners, auxiliary equipment. The industrial robot workstation system integrates industrial robots and peripheral equipment, and uniformly defines the interfaces of each device according to the requirements on the hardware to meet the communication requirements, integrates the information flow of the entire system on the software, and then controls each device according to the process running.

The spot welding robot workstation is shown in Fig. 12-6.

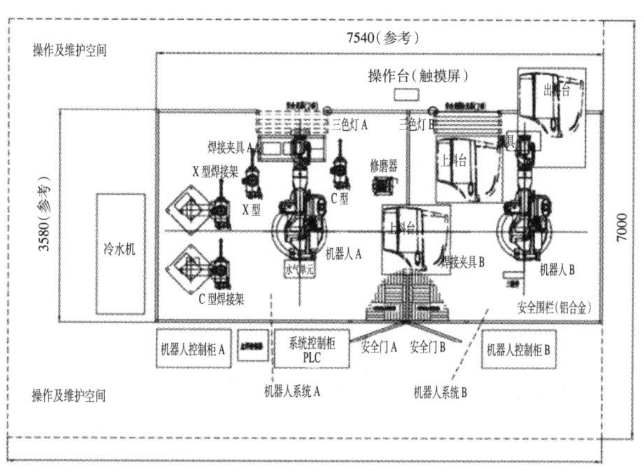

Fig. 12-6 The spot welding robot workstation

图 12-6 点焊机器人工作站

The workstation system includes the following parts: two IRB1410 industrial robots; one ST21 spot welding controller produced by Japan's Ohara Company (Nanjing) Company, which is used to control the spot welding operation and can set the welding current according to the different workpieces; a welding gun cleaner, when the electrode tip of the welding gun is oxidized, it needs to be automatically polished; a set of quick-change device produced by Japan's NITTA company and a set of protective covers for X-shaped welding guns and C-shaped welding guns, quick-change device (installed on the 6th axis flange of the robot) realizes the replacement of the welding tongs, and the welding tongs protective cover (installed on the welding tongs) is used to protect the welding tongs and prevent splash damage during the welding process; one water chiller and an air compressor, the former is used to cool the welding tongs, and the latter supplies air to the entire workstation; a peripheral control cabinet (built-in Siemens S7-300 series PLC), ET200S distributed I/O, which is the command center of the entire control system, when the robot is under external control, the PLC sends instructions; an operation console with a touch screen (Siemens TP270-10) is used to detect the status of the system; a set of water and air unit systems (equipped with electrical proportional valves, flow meters, manual drain valves, etc.), used to detect whether the water and gas of the workstation are ready and meet the requirements; loading platform, unloading platform and other auxiliary peripheral equipment; safety grating and safety door, as important safety

点焊机器人工作站系统包含以下几部分：两台 IRB1410 工业机器人；一台日本小原公司（南京）公司生产的 ST21 型点焊控制器，用来控制点焊操作，可根据加工工件的不同进行电焊电流的设定；一台焊枪清理器，当焊钳电极头氧化后需要自动进行修磨；一套日本 NITTA 公司生产的快换装置及 X 形焊钳和 C 形焊钳的保护罩各一套，快换装置（已安装在机器人第 6 轴法兰盘上）实现焊钳的置换，焊钳保护罩（已安装在焊钳上）用来保护焊钳，防止焊接过程中的飞溅损伤焊钳；一台冷水机和一台空气压缩机，冷水机用来冷却焊钳，空气压缩机为整个工作站供气；一台周边控制柜（内置西门子 S7-300 系列 PLC）、ET 200S 分布式 I/O，这是整个控制系统的指挥中心，当机器人处于外部控制时，由 PLC 发送指令；一台带有触摸屏（西门子 TP270-10）的操作台，用于检测系统状态；一套水气单元系统（内装有电器比例阀、流量计、手动排水阀等），用来检测工作站的水、气是否准备好，并检查其是否符合要求；上料台、下料台及其他附属周边设备；安全光栅及安全门，作为系统重要的安全防护。

周边设备的控制及车门的装夹、搬运和焊接过程的控制逻辑由配套的机器人控制器内的西门子集成通信处理器（CP5614）、可编程序控制器（PLC）和用户焊接示教程序来共同完成。

protection of the system.

The control logic of the peripheral equipment and the clamping, handling and welding process of the car door are carried out together by the Siemens integrated communication processor (CP5614), programmable logic controller (PLC) and user welding teaching program in the supporting robot controller.

微信扫一扫

教学视频

Lecture 12-2
Knowledge of Industrial Robot Workstations (2)

12-2 了解机器人工作站（2）

1.Arc Welding Workstation

The arc welding workstation system is shown in Fig. 12-7.

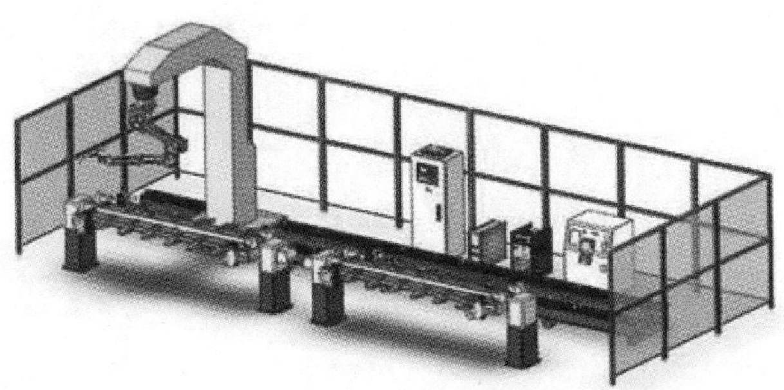

Fig. 12-7 Arc Welding Workstation System

图 12-7 弧焊工作站系统

The arc welding workstation welds the weld between the blade and the teeth according to the process requirements of the workpiece drawing. The welding object is shown in Fig. 12-8. The workpiece is made of carbon steel and the maximum size is 2900mm×540mm×110mm. The welding process requires that the position error of each plate and tooth does not exceed 10mm, and the weld is not allowed to have undercuts, pores, sand holes, cracks and other defects. Taking the weld with a weld height of 12mm and a weld length of 1000mm as a reference, welding is required 6 times with the welding speed of 450mm/min (two passes are completed by three times), the

1. 弧焊机器人工作站

弧焊工作站系统如图 12-7 所示。

该弧焊工作站按照工件图工艺要求焊接刀板和刀齿之间的焊缝，焊接对象如图 12-8 所示，工件材质是碳钢，最大尺寸为 2900mm×540mm×110mm。焊接过程要求各板和齿的位置误差不超过 10mm，焊缝不允许存在咬边、气孔、砂眼、裂纹等缺陷。以焊高为 12mm，焊缝长度为 1000mm 的焊缝为参考，要求焊接 6 次，焊接速度为 450mm/min（两道三次焊完），焊接时间大约 13.4min ，变位机空行程及机器人行走时间为 5min。因为本系统为双工位，工件在焊接的同时就可完成工件装夹，所以工件装夹辅助时间为 0，焊完工件总耗时约 18.4min。

welding time about 13.4min, and the idle stroke of the positioner and the robot walking time are 5min. Because this system is a double station, the workpiece can be clamped at the same time as the workpiece is welded, so the auxiliary time of the workpiece clamping is 0, and the total time for welding the workpiece is about 18.4min.

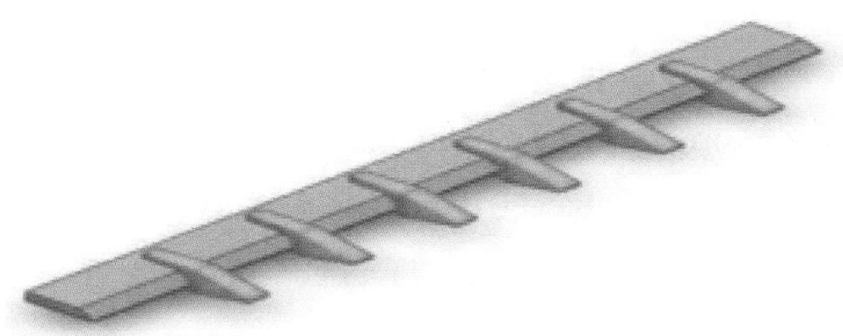

Fig. 12-8 Welding Object
图 12-8 焊接对象

（1）弧焊机器人工作站的工作流程：

①人工通过天车将已组对好的工件吊入变位机工装上并定位加紧，如图 12-9 所示；

(1) The workflow of the arc welding robot workstation:

① Manually hoist the assembled workpiece into the positioner tooling through the crane and tighten the positioning, as shown in Fig. 12-9;

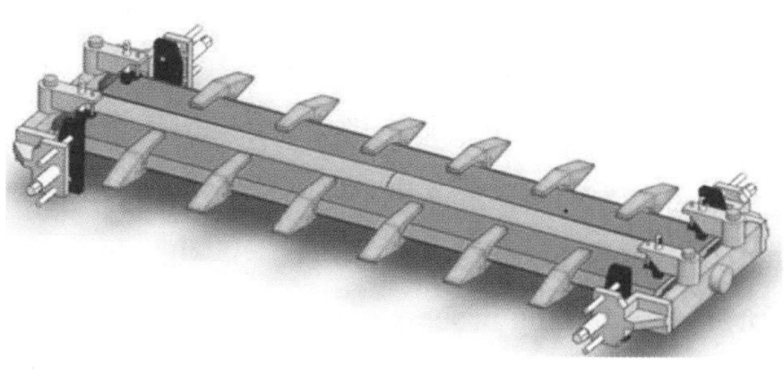

Fig. 12-9 Arc Welding Workstation System
图 12-9 弧焊工作站系统

②机器人从安全区域移动到工作区域，准备对工件进行焊接；

③机器人对工件进行焊接：当机器人对工件 A 的工件焊接完成后，通过机器人行走

②The robot moves from the safe area to the work area, ready to weld the workpiece;

③The robot welds the workpiece: When the robot finishes welding the workpiece of workpiece A, it moves to station B through the robot walking device (while welding, the assembly of station B has been manually completed), and the workpiece on station B is welded. The double station alternates until the end of welding.

(2)The structure of the arc welding robot workstation:

①Robot system

industrial robot system FANUC M-101A.

②Welding system

The high-performance digital welding power source produced by Suzhou Migatronic is selected, the model is Sigma 500 double pulse type. This power source can be combined with a dedicated wire feeder for robots and a water-cooled robot welding gun to form a welding system.

③Water-cooled welding gun

German Tbi RM80W products are chosen as the water-cooled welding gun. Innovatively adopts a two-way shielding gas design. The shielding gas of the outer channel forms an axial shielding air flow, while the shielding air of the inner channel forms a radial shielding air flow. Because the flow control of the shielding gas is strengthened, the laminar shielding gas is gently wrapped in the welding area, the gas is not easy to lose, and the amount of shielding gas during the welding process is greatly reduced. At the same time, the two-channel shielding gas can reach multiple parts in the welding torch, and the cooling effect of the welding torch is enhanced. because of the radial airflow, the contact tip is also fully cooled, which can avoid excessive burning of the contact tip. As a result, the life span of the contact tip is extended by 3 to 5 times.

装置移动到工位 B（在焊接的同时，工位B已由人工完成装件），对工位 B 上的工件进行焊接。双工位交替进行，直到焊接结束。

（2）弧焊机器人工作站的结构

①机器人系统

工业机器人系统 FANUC M-101A。

②焊接系统

选用苏州米加尼克公司生产的高性能数字化焊接电源，型号为 Sigma500 双脉冲型。这种电源可以和机器人专用送丝机及水冷机器人焊枪组成焊接系统。

③水冷焊枪

水冷焊枪选用德国 Tbi RM80W 产品。创新地采用两路保护气设计，外层通道的保护气形成轴向的保护气流，而内层通道的保护气形成径向保护气流，因为加强了保护气的流动控制，使得层流状保护气在焊接区温柔包裹，气体不易散失，焊接过程中的保护气用量大大降低。同时，两通道保护气可以到达焊枪内多个部位，焊枪的冷却效果得到加强。因为径向气流的存在，使得导电嘴也得到充分地冷却，可避免导电嘴的过度烧损，从而使导电嘴的寿命延长了 3~5 倍。

④清枪装置

清枪装置选用 Tbi 焊枪清理装置，型号为 BRG2000。焊枪清理单元是每一个 MIG/MAG 焊接机器人不可缺少的部分，机器人只需要执行一个动作，就可以利用两个旋转的刀片清除掉焊枪内部附着的黏着物，并喷射硅油，确保焊渣和喷嘴不会发生死粘连。清理的频率取决于不同的条件，如工艺过程、气体类型、材料类型等。程序员在机器人控制器的工作程序中确定清理程序的频率。采用清枪装置能明显减少中断时间，延长焊枪部件的使用寿命且不需要手工操作。

⑤单轴精密变位机

焊接变位机是焊接夹具的载体，能使工件达到最佳的焊接位置。本套焊接系统配置的变位机由变位机头座、变位机尾座、专用工装夹具等部分组成。

⑥焊接夹具

本套焊接夹具主要由 L 支撑板、侧向定位装置、侧向压紧装置、预变形装置、上端旋转压紧装置等部分组成。

⑦机器人工作站的中央程控系统

程序控制系统主要用于控制机器人、变位机和其他相关

④Gun cleaning device

The gun cleaning device adopts the Tbi welding gun cleaning device, and the model is BRG2000. The cleaning unit of welding gun is an indispensable part of each MIG/MAG welding robot. The robot only needs to perform one action, and it can use two rotating blades to remove the adhesion inside the welding gun, and spray silicone oil to ensure that the welding slag and nozzle will not stick together. The frequency of cleaning depends on different conditions, such as process, gas type, material type, etc. The programmer determines the frequency of the cleaning program in the working program of the robot controller. The use of the gun cleaning device can significantly reduce the interruption time, extend the service life of the welding torch components,and does not require manual operation.

⑤Single axis precision positioner

The welding positioner is the carrier of the welding fixture, which can make the workpiece reach the best welding position. The positioner of this set of welding system is composed of positioner headstock, positioner tailstock, special fixtures and other parts.

⑥Welding fixture

This set of welding fixture is mainly composed of L support plate, lateral positioning device, lateral compression device, pre-deformation device, upper end rotary compression device and other parts.

⑦Central program control system of robot workstation

The program control system is mainly used to control the orderly movement of other related parts such as robots and positioners. The program control system in the workstation adopts a variable program control system, and its main control part adopts PLC as the main control unit. The system has the function of "manual" and "automatic" option. In the "manual" mode, manual participation is possible, and in the "automatic" mode, the corresponding tasks are automatically completed in each area. The system is equipped with: power on/off and indication buttons, emergency stop button, which can be stopped in an emergency when an accident occurs. Other electrical components of the control system include relays, switches and buttons. The robot is the core of the entire system. The system can conveniently set the welding process and parameters (welding program) for each type of workpiece. The welding program can be stored and called at any time. The automatic welding of the workpiece is completed according to the welding program selected by the operator during work.

部分有序运动。工作站中的程序控制系统采用可变程序控制系统，其主控部分采用 PLC 作为主控单元，系统具有“手动”“自动”选择功能，在“手动”模式下可以人工参与，在“自动”模式下各区域自动完成相应工作。系统中设有电源开/关及指示按钮，急停按钮，当发生意外时可紧急停止。控制系统的其他电气元件还包括继电器、开关和按钮等。机器人是整个系统的核心，该系统可以对每种工件都方便地设定焊接工艺及参数（焊接程序），焊接程序可进行存储并被随时调用，工作时按操作者选用的焊接程序完成工件的自动焊接。

2.stamping robot workstation

A typical robotic stamping automation production line includes the following parts: stamping robots, control systems, destacking systems, retractable belt conveyors, sheet cleaning machines, sheet lubricating machines, sheet centering stations, seven-axis robots and line tail stacking system. The specific layout can be adjusted according to the area of the production workshop. For example, the way the destacker can be pulled out can be parallel to the punching line or perpendicular to the punching line.

2. 冲压机器人工作站

典型的机器人冲压自动化生产线包含以下部分：冲压机器人、控制系统、拆垛系统、可伸缩皮带机、板料清洗机、板料涂油机、板料对中台、机器人七轴及线尾码垛系统。具体布置方式可以根据生产车间的面积进行调整，如拆垛车的开出方式既可以和冲压线平行，也可以和冲压线垂直。

（1）冲压机器人

冲压机器人除了要求具有负载大、运行轨迹精确及性能稳定可靠等搬运机器人所共有的特性,还要满足频繁起/制动、作业范围大、工件尺寸及回转面积大等特点。各个厂家的冲压机器人都在普通搬运机器人的基础上加大了电动机功率及减速机规格，加长了手臂，并广泛采用棚架式安装结构。

（2）控制系统

机器人冲压自动化系统需要集成压力机、机器人、拆垛机、清洗机、涂油机、对中台、双料检测装置、视觉识别系统、各种皮带、同步控制系统、安全防护系统及大屏幕显示，并具有无缝集成进工厂 MES 系统的能力。为了把如此多的智能控制系统有效集成，一般采用以太网和工业现场总线二级网络系统，其中现场总线系统能同时搭载安全总线。

（3）拆垛系统

目前常见的拆垛系统有三种，分别是专用拆垛机、机器人+拆垛小车、桁架式机械手+拆垛小车。

①专用拆垛机

其结构特点是垛料放置在可移动液压升降台车上；垛料高度依靠光电传感器和液压系

(1) Stamping robot

The stamping robot not only requires large load, accurate running trajectory, stable and reliable performance and other characteristics common to handling robots, but also meets the characteristics of frequent starting/braking, large working range, large workpiece size and rotating area. The stamping robots of various manufacturers have increased the motor power and reducer specifications on the basis of ordinary handling robots, lengthened the arm, and widely adopted the scaffolding installation structure.

(2) Control system

The robot stamping automation system needs to integrate presses, robots, destackers, washing machines, oiling machines, centering stations, double-material detection devices, visual recognition systems, various belts, synchronization control systems, safety protection systems and large screen displays. It also has the ability to seamlessly integrate into the factory MES system. In order to effectively integrate so many intelligent control systems, Ethernet and industrial fieldbus secondary network systems are generally used, and the fieldbus system may also be equipped with a safety fieldbus.

(3) Destacking system

At present, there are three common destacking systems, which are dedicated destacker, robot + destacking trolley, and truss type manipulator + destacking trolley.

①Dedicated destacker

Its structural feature is that the stack material is placed on a movable hydraulic lifting trolley; the he ight of the stack material is controlled by a photoelectric sensor and a hydraulic system to keep it constant; the magnetic splitter relies on pneumatic or electric drive to automatically close to the stack material; the vacuum chuck group driven by a cylinder and arranged in a matrix is used for destacking, and the vacuum chuck group moves vertically; the disassembled sheets are transported by magnetic belts.

②Robot+destacking trolley

Its structural feature is that the stacking material is placed on a movable destacking trolley; the height of the stacking material is not controlled, and the robot suction height is automatically adjusted by the calculated sheet thickness during destacking; the bracket of the magnetic divider is installed on the destacking trolley. The bracket can be translated and has multiple adjustable joints that can rotate freely. When changing the stacking material, manually place the magnetic divider against the stacking material periphery; the vacuum suction cup group and the double material detection sensor for the stacking are installed on the robot end picker; the disassembled sheets are placed on the telescopic transition belt by the robot for transmission.

③Truss type manipulator+destacking trolley

Its structural features are basically the same as the “robot+destacking trolley”, but the difference is that the robot in the “robot+destacking trolley” is replaced by a truss manipulator to complete each action. Its structural feature is that the stacking material is placed on a movable destacking trolley. The height of the stacking material is not controlled, and the height

统控制，保持恒定；磁力分张器依靠气动或电动驱动自动贴近垛料；采用气缸驱动、矩阵布置的真空吸盘组进行拆垛，真空吸盘组垂直运动；拆成单张的板料采用磁性皮带传输。

②机器人+拆垛小车

其结构特点是垛料放置在可移动的拆垛小车上；垛料高度不控制，拆垛时依靠计算的板料厚度自动调整机器人吸料高度；磁力分张器支架安装在拆垛小车上，支架可平移并具有多个可自由旋转的调整关节，更换垛料时人工将磁力分张器靠在垛料周边；拆垛用真空吸盘组及双料检测传感器安装在机器人端拾器上；拆成单张的板料由机器人放置在可伸缩过渡皮带上进行传输。

③桁架式机械手+拆垛小车

其结构特点与“机器人+拆垛小车”基本相同，不同之处在于用桁架式机械手代替了“机器人+拆垛小车”中的机器人来完成各个动作。其结构特点是垛料放置在可移动的拆垛小车上，垛料高度不控制，拆垛时依靠计算的板料厚度自动调整机械手吸料高度，磁力分张器支架安装在拆垛小车上，支架可平移并具有多个可自由旋转的调整关节，更换垛料时人工将磁力分张器靠在垛料周边；拆垛用真空吸盘组及双料检测传感器安装在桁架式机械手端拾器上：拆成单张的板料由机械手放置在可伸缩过渡皮带上进行传输。

（4）可伸缩皮带机

可伸缩皮带机用于拆垛完成后单张板料的传输，它可以为板料提供穿过清洗机所需的移动速度。皮带机的速度一般为变频控制，以保证与后续清洗机、涂油机的同步。皮带长度可调整，以填补不进行板料清洗、清洗机离线时的空间。

（5）板料清洗机

随着客户对汽车表面质量的要求越来越高，在冲压成型前对覆盖件板料进行清洗已经成为汽车厂的标准工艺。根据产量大小，板料清洗可以分为在线清洗和离线清洗两种工艺：离线清洗适用于投产初期，单车型产量较低的企业，一条板料清洗线可以供应2—3条冲压线，离线清洗机的规格一般宽度为2m；在线清洗适用于批量较大的企业，为适应轿车整体的尺寸，在线清洗机的规格一般宽度为4.2m。

of the manipulator suction is automatically adjusted by the calculated sheet thickness during destacking: the magnetic splitter bracket is installed on the destacking trolley. The bracket can be translated and has multiple adjustable joints that can rotate freely. When changing the stacking material, manually place the magnetic divider against the stacking material periphery; the vacuum suction cup group and the double material detection sensor for the stacking are installed on the truss manipulator end picker; the disassembled sheets are placed on the telescopic transition belt by the manipulator for transmission.

(4) Retractable belt conveyor

The retractable belt conveyor is used to transfer a single sheet after the destacking is completed, and it provide the sheet with the required moving speed through the washing machine. The speed of the belt conveyor is generally controlled by frequency conversion to ensure synchronization with the subsequent washing machine and oiling machine. The length of the belt can be adjusted to fill the space when the sheet metal is not cleaned and the cleaning machine is offline.

(5) Sheet cleaning machine

As customers have higher and higher requirements for the surface quality of automobiles, cleaning the cover sheet before stamping has become a standard process in automobile factories. According to the output size, sheet metal cleaning can be divided into two processes: online cleaning and offline cleaning: offline cleaning is suitable for enterprises with low output of single vehicle models at the early stage of production, and one sheet cleaning line supply 2-3 stamping lines. The general width of the machine is 2m; the online cleaning is suitable for enterprises with larger batches. In order to adapt to the size of the car, the general width of the online cleaning machine is 4.2m.

The cleaning machine is composed of a roller system consisting of a guide roller, a brush roller and a squeezing roller, a power and transmission system, a hydraulic adjustment mechanism, a cleaning filter system, a cluster nozzle assembly, an oil mist collector, a lubrication system, a walking mechanism, an electrical control system and other components. The washing machine is mainly used for the cleaning of standard sheets, uncoated, galvanized sheets and aluminum sheets and coils. The washing machine has a self-propelled mechanism, and the whole machine can be driven out along the ground track when the washing process is not required.

The guide roller, brush roller, and squeezing roller each have independent pressure adjustment and frequency conversion drive mechanism. By accurately adjusting the pressure and speed, it can ensure that the worn and repaired roller system is synchronized with the production line. The squeezing roller adopts a non-woven laminated cloth roller to provide good squeezing and tensioning performance, and has anti-scratch and self-healing functions. The cleaning oil tank has a heating system, which can make the cleaning medium be used under different ambient temperatures and achieve the best oiling effect. The electrical system adopts PLC control with

清洗机由引料辊、毛刷辊及挤干辊组合的辊系，动力及传动系统、液压调整机构，清洗过滤系统，集束喷管组件，油雾收集器，润滑系统，行走机构，以及电气控制系统等组成。主要用于标准板材、未涂层镀锌板以及铝板等板料、卷料的清洗。清洗机具有自行走机构，在不需要清洗工序时，整机可沿地面轨道开出。

引料辊、毛刷辊、挤干辊各自具有独立的压力调整与变频驱动机构，通过精确调整压力与速度，可以保证磨损、维修后的辊系与生产线保持同步，挤干辊采用无纺布层压布辊，以提供良好的挤干和张紧性能，并具有防擦伤和自愈功能。清洗油箱具有加热系统，可以使清洗介质在不同环境温度下使用并达到最佳涂油效果。电气系统采用具有现场总线通信功能的 PLC 控制，通过触摸屏人机界面可以进行参数设定、故障诊断。

（6）板料涂油机

为了保证板料在较高速度下拉延成型时的质量，在冲压成型前对板料进行局部涂抹拉延油是汽车厂常用的工艺手段，适用于自动化冲压线的涂油机。

涂油机主要用于板材拉延前的涂油工序，由喷枪单元、供油及油液保温单元、喷射供气单元、板料输送单元、油雾收集单元、电气控制单元及机身等组成，涂油机具有行走机构，在不需要涂油时可以移动离线。喷嘴采用数字化控制，保证板料上油膜位置的准确和油膜厚度的均匀。不管环境温度如何变化，供油及油液保温单元均处于等待喷射状态，油液温度恒定，保证喷射效果。

（7）板料对中台

为了保证板料被准确地放置在模具内，在机器人拾取板料前需要对板料进行对中。常用的对中台有重力对中台、机械对中台及光学对中台三种。

①重力对中台

板料在装满滚珠的斜面上依靠重力自由滑向一角进行定位，台上具有板料到位检测及双料检测。重力对中台适用于

fieldbus communication function, and parameter setting and fault diagnosis can be carried out through the touch screen man-machine interface.

(6) Sheet oiling machine

In order to ensure the quality of the sheet metal when it is drawn and formed at a higher speed, partial application of drawing oil to the sheet metal before stamping is a commonly used process method in automobile factories, which is suitable for the oiling machine of the automatic press line.

The oiling machine is mainly used for the oiling process before the drawing of the sheet. It consists of a spray gun unit, an oil supply and oil heat preservation unit, an injection air supply unit, a sheet material conveying unit, an oil mist collection unit, an electrical control unit and a body. The oiling machine has a walking mechanism, which can be moved offline when no oiling is needed. The nozzle adopts digital control to ensure the accurate position of the oil film on the sheet and the uniform thickness of the oil film. No matter how the ambient temperature changes, the oil supply and oil heat preservation units are in the waiting state for injection, and the oil temperature is constant to ensure the injection effect.

(7) Sheet centering station

In order to ensure that the sheet material is accurately placed in the mold, the sheet material needs to be centered before the robot picks up the sheet material. Commonly used centering stations include gravity centering stations, mechanical centering stations and optical centering stations.

①Gravity centering station

The sheet material is positioned on a slope filled with balls by freely sliding to a corner by gravity, and the table has sheet material in-position detection and double-material detection. The gravity centering station is suitable for robots to directly place sheets, but not for automatic lines with washing machines and oiling machines.

②Mechanical centering station

The sheet material is moved to the stop by the magnetic leather bag, and the other three cylinders drive the feeder to the center to achieve precise positioning. The position of all the feeders can be taught and programmed, and the centering station has sheet material in-position detection and double-material detection. The mechanical centering station can meet the high-speed centering of various heterosexual materials. In addition, a dual-center design can be used to meet the requirements of processing one mold with two parts or dual molds with two parts at the same time. The station has sheet material in-position detection and double material detection.

③Optical centering station

This is a technology developed in recent years. Its principle is to automatically adjust the trajectory of the robot after the image of the position of the sheet obtained by taking pictures through the video processing software, so as to not only meet the requirements of accurately placing the sheet into the mold, but also avoid the complex organization of the mechanical centering station. The optical centering station is essentially a magnetic belt conveyor with

机器人直接放置板料，不适用于具有清洗机、涂油机的自动线。

②机械对中台

板料依靠磁性皮袋移动至挡块，再由另外三侧的气缸驱动打料器向中心推从而实现精确定位。所有打料器的位置可示教编程，对中台具有板料到位检测及双料检测。机械对中台能够满足各种异性料的高速对中。另外，可采用双中心设计，以满足同时处理一模两件或双模双件的要求。台上具有板料到位检测及双料检测。

③光学对中台

这是最近几年发展的技术，其原理是利用拍照获得的板料位置图像经视频处理软件后自动调整机器人的轨迹，从而既满足最终将板料精确置入模具内的要求，又避免了机械对中台的复杂机构。光学对中台本质上是一个具有板料到位检测及双料检测的磁性皮带机，相比机械对中台，光学对中台能够显著降低成本。

（8）机器人七轴与端拾器

标准六轴机器人用于冲压生产能够达到 8 件/min 的生产节拍，为了进一步提高生产率，近年来机器人制造厂家与系统集成商纷纷研发附加七轴。采用七轴后，机器人冲压自动化生产线的生产节拍可以进一步提高到 12 件/min，使得机器人自动化冲压线的生产率能够与昂贵的专用机械手高速线一较高下。

（9）线尾码垛系统

自动化冲压线的生产节拍能够超过 10 件/min，这时线尾码垛工位的负荷就超过了劳动极限。为了适应这种高节奏的产出，需要先对冲压件成品进行分流，然后由人工或机器人将其码入料架，最后由铲车将料架运走。

sheet metal in-position detection and double-material detection. Compared with the mechanical centering station, the optical centering station can significantly reduce the cost.

(8) Robot seven - axis and end picker

Standard six-axis robots can achieve a production cycle of 8 pieces/min for stamping production. In order to further improve productivity, robot manufacturers and system integrators have developed additional seven-axis in recent years. After adopting the seven-axis, the production cycle of the robot stamping automation production line can be further increased to 12 pieces/min, so that the productivity of the robot automation stamping line can be compared with the expensive special manipulator high-speed line.

(9) Line tail stacking system

The production cycle of the automated stamping line can exceed 10 pieces/min. At this time, the load of the stacking station at the end of the line exceeds the labor limit. In order to adapt to this kind of high-tempo output, the finished stamping parts need to be diverted first, and then manually or robotically packed into the rack, and finally the rack is transported away by a forklift.

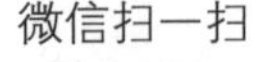

教学视频

Lecture 13-1

CNC Machine Tool Industrial Robot Workstation(1)

13-1 数控机床机器人工作站（1）

The present manufacturing industry is transforming from traditional to intelligent manufacturing, so intelligent manufacturing equipment and technology have attracted more and more attention. The cutting intelligent manufacturing unit is based on the related advanced technology, using modern network communication technology, sensor technology, detection technology, automation technology, anthropomorphic intelligent technology and other intelligent manufacturing technology through information technology and intelligent technology, finally realized the unmanned automation unit.

1. the role of cutting manufacturing unit

The cutting manufacturing unit is an intelligent manufacturing unit used to process cylinder sleeves. This intelligent manufacturing unit uses a PLC master control system to control machine tools, industrial robots, and 2D visual recognition to realize the entire automated production.

This set of CNC machine tool industrial robot workstations can realize unmanned loading and unloading, material position identification, and various machining processes, such as material

现在的制造行业正从传统的制造向智能制造转变，智能制造装备和技术越来越受到大家的关注。切削加工智能制造单元就是以智能制造单元相关先进技术为基础，运用了现代网络通信技术、传感技术、检测技术、自动化技术、拟人化智能技术等智能制造技术，通过信息技术和智能化技术，最终实现了无人化自动化单元。

1. 切削加工制造单元的作用

切削加工制造单元是用于加工油缸套筒的智能制造单元，本智能制造单元通过PLC主控系统，控制机床、工业机器人、2D视觉识别等实现整个自动化生产。

这套数控机床工业机器人工作站可以实现无人化的上、下料，物料位置的识别，以及多种机械加工过程，如物料的车削、铣削、钻削、孔口的导角和攻丝。

通过这套机器人工作站可以学习到最先进的智能制造技术，例如工业机器人的编程和操作、数控车床和数控加工中心的编程和操作、机器视觉的操作、MES 执行系统的操作，以及智能设备之间的组网通信技术。

turning, milling, drilling, orifice lead angle and tapping.

Through this robot workstation can learn the most advanced intelligent manufacturing technology, such as the programming and operation of industrial robots, the programming and operation of CNC lathes and CNC machining centers, the operation of machine vision, and the operation of MES execution systems, and network communication technology among smart devices.

2. 切削加工制造单元的结构

切削加工制造单元包含的设备有：主控系统、数控车床、立式加工中心、固定机器人、2D 视觉系统、清洁装置、上下料接驳料架、监控操作台、翻转机构、电气柜以及安全护栏。切削加工制造单元的结构如图 13-1 所示。

2. the structure of cutting manufacturing unit

It includes main control system, CNC lathe, vertical machining center, fixed robot, 2D vision system, purging device, loading and unloading feeder rack, monitoring console, turning mechanism, electrical cabinet and safety guardrail.

The structure of cutting manufacturing unit is shown in Fig. 13-1.

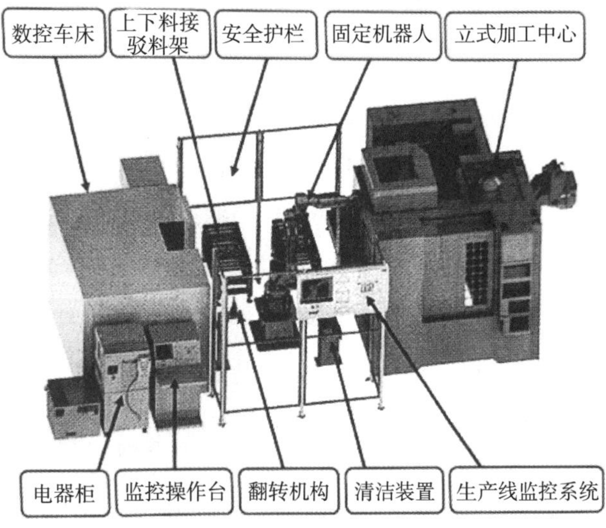

Fig. 13-1 the structure of cutting manufacturing unit

图 13-1 切削加工制造单元

(1) Production equipment of robots workstation

In order to realize the processing of cylinder sleeve workpieces, the cutting processing manufacturing unit uses FANUC industrial robots to realize the turnover. loading and unloading of workpieces. CNC lathes and vertical machining centers guarantee the processing and production of cylinder sleeve workpieces.The robot workstationis shown in Fig. 13-2.

（1）机器人工作站的生产设备

为实现油缸套筒工件的加工，切削加工制造单元使用FANUC 工业机器人实现工件的周转和机床的上下料动作；数控车床和立式加工中心保障了油缸套筒工件的加工生产。机器人工作站如图 13-2 所示。

Fig. 13-2 production equipment of robot workstation

图 13-2 机器人工作站的生产设备

①CNC Lathe

The CNC lathe of the robot workstation adopts the NL201HA rolling guide type CNC horizontal lathe of FANUC 0i-TF PLUS CNC system. Overall bed, and parallel plastic guide rails, high rigidity and easy chip removal. Equipped with a high-precision spindle with small runout, the maximum spindle speed is 6000 rpm. The tool post is a hydraulic tool post with stable operation, fast indexing speed and high reliability. Realizing the external lead angle machining of cylinder sleeves workplace.CNC lathe is shown in Fig. 13-3.

①数控车床

该机器人工作站的数控车床采用 FANUC 0i-TF PLUS 数控系统的 NL201HA 滚动导轨型数控卧式车床。整体的床身，并行贴塑导轨，高刚性，易排屑。配备高精度主轴，跳动小，主轴最高转速 6000rpm。刀架为液压刀架，工作平稳、转位速度快、可靠性高。实现油缸套筒工件的外圆导角加工。数控车床如图 13-3 所示。

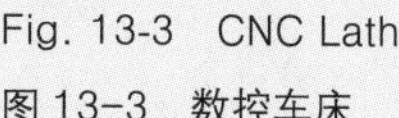

Fig. 13-3 CNC Lathe

图 13-3 数控车床

②立式加工中心

立式加工中心采用 FANUC 0i-MF PLUS 数控系统的高效型立式加工中心。配备高速主轴单元,主轴测试温升小、热变形小、加工精度高。高精度丝杠,长寿命轴承,重切削、高速切削导轨。刀库为嵌入"卡刀一键复原功能",有效提高刀库故障解除效率。大功率、大扭矩主轴电机,可选配德国进口的 ZF 减速箱,增加输出转矩。立式加工中心还配备第四转台。加工中心可以实现油缸套筒工件的钻底孔、攻螺纹和铣侧面加工。立式加工中心如图 13-4 所示。

②Vertical machining center

The vertical machining center adopts the high-efficiency vertical machining center of FANUC 0i-MF PLUS CNC system. Equipped with high-speed spindle unit, spindle test temperature rise is small, so thermal deformation is also low, but processing accuracy is high. It adopteds high-precision lead screw, long-life bearing, heavy cutting, high-speed cutting guide. The tool magazine is embedded with the "one-key recovery function for stuck tools", which effectively improves the efficiency of trouble-shooting the tool magazine. High-power, high-torque spindle motor, ZF gearbox imported from Germany can be selected to increase output torque. The vertical machining center is also equipped with a fourth turntable. The machining center can realize bottom hole drilling, thread tapping and side milling of cylinder sleeve workpieces. Vertical machining centeris shown in Fig. 13-4.

Fig. 13-4 vertical machining center
图 13-4 立式加工中心

(2) Fixed robot

The fixed robot adopts FANUC's M-20iD25 fixed articulated robot, and the total number of control axes is 6. It is installed in a fixed position of the cutting processing unit, and the robot's two-handed claws are used for workpiece picking and loading and unloading with high positioning accuracy and grasping stability. It can realize the loading and unloading of the machine tool for the entire unit system.

(3) 2D vision system

The vision system consists of a 2D camera mounted on the gripper of an industrial robot to complete visual data collection. The vision system is used as a positioning method for accurate grasping of the workpiece to be processed, eliminating the need for mechanical pre-positioning fixtures that must usually be used to satisfy the accurate grasping of robots, and has high flexibility. It is mainly through the visual system software setting to establish the

（2）固定机器人

固定机器人采用 FANUC 的 M-20iD25 固定关节型机器人，总控轴数为 6 轴。它安装在切削加工单元的固定位置，机器人双手爪用于工件的拾取及上下料，有较高的定位精度和抓持稳定性。它可以实现整个单元系统的机床上下料动作。

（3）2D 视觉系统

该视觉系统由一个安装于工业机器人手爪上的 2D 摄像头完成视觉数据采集。该视觉系统作为待加工工件准确抓取的定位方式，省去通常为满足机器人的准确抓取而必须采用的机械预定位夹具，具有很高的柔性。主要是通过视觉系统软件设置，建立视觉画面上的点位与机器人位置相对应关系。同时对工件进行视觉成像并与已标定的工件进行比较，得出偏差值，即机器人抓放位置的补偿值，通过补偿实现机器人自动抓放，可实现机器人对无夹具定位工件的自动柔性搬运。2D 视觉系统如图 13-5 所示。

corresponding relationship between the points on the visual screen and the robot position. At the same time, the workpiece is visually imaged and compared with the calibrated workpiece, and the deviation value is obtained, that is, the compensation value of the robot's pick-and-place position. Through compensation, the robot can automatically pick and place, which can realize the robot's automatic and flexible handling of the workpiece without fixture. The 2D vision system is shown in Fig. 13-5.

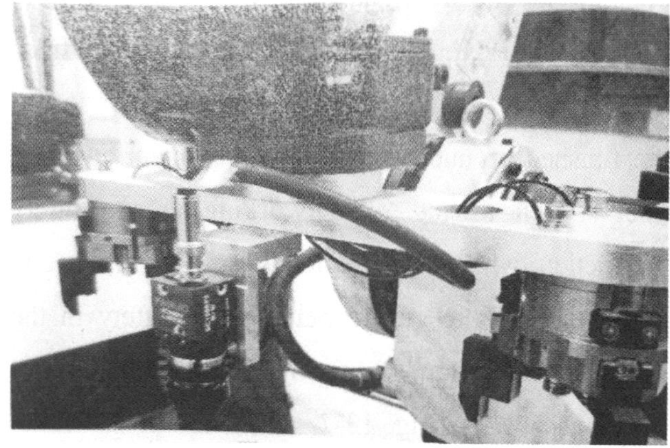

Fig. 13-5 2D machine vision system
图 13-5 二维视觉系统

（4）清洁装置

清洁装置实现对工件的自动吹气清洁。如图 13-6 所示。

(4) purging device

The purging device realizes automatic blowing and cleaning of the workpiece.As shown in Fig. 13-6 .

Fig. 13-6 purging device
图 13-6 清洁装置

(5) Main control system

The main control system adopts Siemens S7-1200 PLC controller and uses the man-machine interface to monitor and control the running status of entire system, as well as realizes the transmission of real-time and non-real-time data in the system with high reliability and maintainability. The main control system is loaded with MES production execution system, optimized the production process through information interaction with the upper business planning layer (ERP) and the lower process control layer (CNC, PLC, Robot) within the time range from the start of receiving the order to the completion of the finished product. The safety equipment adopts a door switch as the robot working area's safety production, which can completely isolate workers and machine, so as to ensure personnel's safety in the automatic operation of this system. Main control systemis shown in Fig. 13-7.

（5）主控系统

主控系统采用西门子S7-1200PLC控制器，运用人机界面对整个系统的运行状态进行监控，实现系统中实时和非实时数据的传输，具有高度可靠性和可维护性。主控系统装载了MES生产执行系统，在接收订单开始到完成成品的时间范围内，通过与上层业务计划层（ERP）和底层过程控制层（CNC、PLC、Robot）进行信息交互，实现生产过程的优化。安全设备采用门开关，作为机器人工作区域的安全防护，完全做到人机隔离，确保系统在自动运行中的人员安全。主控系统如图13-7所示。

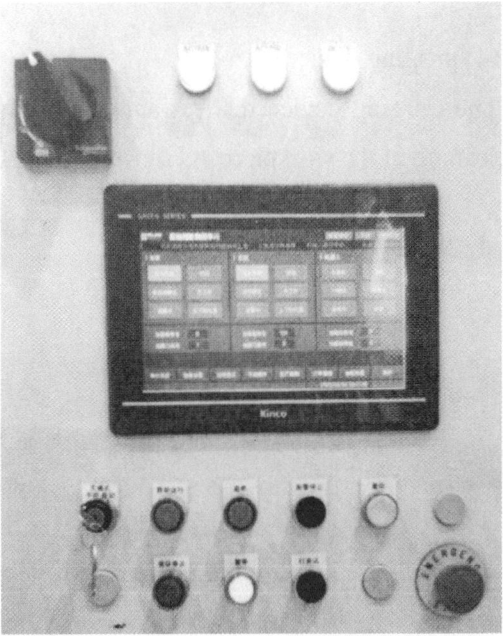

Figure. 13-7 main control system

图13-7 主控系统

3. 切削制造单元的加工过程

首先，AGV 小车装载着待加工件，并把物料放到上料接驳料架上，然后机器人手爪上方的 2D 视觉识别零件整体位置，工业机器人视觉识别工件 1，手爪 A 抓取工件 1，运行至数控机床外侧，工业机器人送工件至数控车床进行 OP10 工序，然后机器人将工件取出，放在翻转台上翻转工件，再把工件送入数控车床进行 OP20 工序，加工完成后，机器人把工件放入立式加工中心进行 OP30 工序，然后机器人把工件放在清洁装置进行清洁，然后把工件放在翻转台上翻转，再放在清洁装置进行清洁，清洁完成后，机器人把物料放在下料接驳料架上，最后由 AGV 小车运走。

3. The machining process of cutting manufacturing units

Firstly, the AGV is loaded with the workpiece to be processed, and the material is placed on the loading feeder rack, then integral position of parts is identified by 2D visual recognition above the robot claw. The industrial robot visually recognizes the workpiece 1, and the hand grasps A to grab the workpiece 1 , Run to the outside of the CNC machine tool, the industrial robot sends the workpiece to the CNC lathe for the OP10 process, then the robot takes the workpiece out, puts the workpiece on the turning table, and turns the workpiece, and then sends the workpiece to the CNC lathe for the OP20 process. After the processing is completed, the robot takes the workpiece, and put it into the vertical machining center for the OP30 process, then the robot puts the workpiece on the cleaning device for cleaning, next puts the workpiece on the turning table, and then places it on the cleaning device for cleaning. After the cleaning is completed, the robot puts the material on the blanking feeder rack, and finally transported away by the AGV.

Lecture 13-2
CNC Machine Tool Industrial Robot Workstation(2)

微信扫一扫

教学视频

13-2 数控机床机器人工作站（2）

1. the role of the intelligent manufacturing integrated production line

Based on the college's industry background, major settings and teaching needs, combined with regional industries, the connotation design of the integrated intelligent manufacturing production line is a customized scheme. It is carried out to meet the needs of students' training while linking to the development of the industry, enabling students to fully understand the reality of modern enterprises, as well as ensuring the precise matching of talent training and job requirements.

Intelligent manufacturing integrated production line selects typical industrial parts—flanges and rocket components as processed products. The processing and assembly truly reflects the process flow. Advanced manufacturing technologies such as intelligent processing, online inspection, intelligent warehousing, intelligent logistics AGV, and intelligent production management are applied throughout the production process. At the same time, the integrated manufacturing production line is not only a simple combination of modular intelligent

1. 智能制造综合生产线的作用

智能制造综合生产线的内涵设计，根据学校行业背景、专业设置和教学需求，结合区域产业，进行定制化方案设计，满足学生实训的同时，对接产业发展，能够让学生充分了解现代企业的实际生产过程，也能保证人才培养与岗位需求的精准匹配。

智能制造综合生产线选取典型产业零件——法兰盘和火箭部件为加工产品，产品的加工和组装过程真实体现了工艺流程，整个生产过程中应用了智能加工、在线检测、智能仓储、智能物流 AGV、智能化生产管理等先进制造技术。同时，综合制造生产线不仅是模块化智能单元的简单叠加组合，更是符合智能工厂生产实际，按照产品工艺流程详细布局，满足学生对于智能制造综合生产线的联调联试的训练需求，提升学生的综合技术能力。

2. 智能制造综合生产线的结构

智能制造综合生产线主要包含：3D 散堆识别、立体料库、2 台数控车床和 2 台立式加工中心、机器人、上下料接驳料架、物料缓存台、检测单元、AGV 物料小车、机器人夹具、清洁装置、打标机、装配单元、生产线监控系统以及控制系统。其结构如图 13-8 所示。每个加工设备上均安装有摄像头，可将视频连接到软件平台上进行教学展示和监控。

units, but also in line with the actual production of smart factories. It is laid out in accordance with the detailed product process flow to meet the training needs of students for the joint debugging and joint test of the intelligent manufacturing integrated production line and to improve comprehensive technical ability of students.

2. the structure of the intelligent manufacturing integrated production line

The intelligent manufacturing integrated production line mainly includes: 3D bulk pile recognition system, stereoscopic material warehouse, 2 CNC lathes and 2 vertical machining centers (VMC), robots, loading and unloading convey racks, material buffering tables, inspection units, AGV, robot fixtures, purging device, laser engraving machine, assembly unit, production line monitoring system and control system. It'sstructureas shown in Fig. 13-8. Each processing equipment is equipped with a camera, which can connect the video to the software platform for teaching demonstration and monitoring.

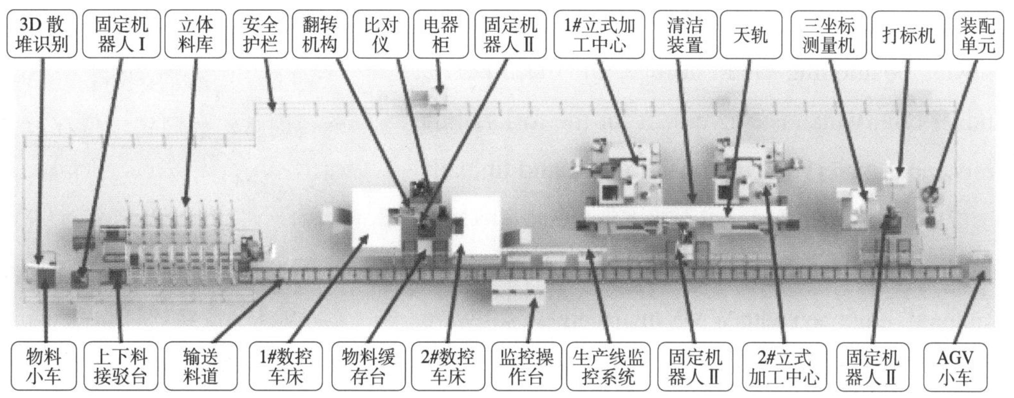

Fig 13-8 the structure of the intelligent manufacturing integrated production line.

图 13-8 智能制造综合生产线的布局图

(1)AGV: transport materials to a fixed track on the ground to realize material sorting by the robot;

(2)Fixed robot 1: installed in a fixed position to realize automatic loading and unloading of materials in the stereoscopic material warehouse;

(3)Fixed robot gripper (workpiece recognition): used by robots to pick parts from scattered workpieces identified by 3D area sensors. Robot grippers have high positioning accuracy and holding stability, high positioning accuracy, good durability, and easy maintenance;

(4)2D vision system: this vision system completes visual data collection by a 2D camera. The vision system is used as a positioning method for accurate grasping of workpieces to be processed, eliminating the need for mechanical pre-positioning fixtures that are usually necessary for accurate grasping by buying robots, and has high flexibility;

(5)3D bulk recognition: this vision system uses a 3D wide-area sensor installed on the ground to detect the position of the workpiece, and then the robot picks it up in bulk;

(6)Loading and unloading conveyor: realizes the turnover positioning and storage of the material turnover box, which is convenient for the forks of the stereoscopic material warehouse to load and unload;

(7)Stereoscopic material warehouse: used to store workpieces to be processed and after processing is completed, equipped with WMS management software, students can learn to master the control of the stereoscopic material warehouse out of the library;

(8)Material conveying channel: realize the conveying

（1）AGV 物料小车：运送物料至地面固定轨道处，实现机器人的物料分拣；

（2）固定机器人 1：安装于固定位置，实现立体料库的物料自动上下料；

（3）固定机器人手爪（工件识别）：用于机器人实现从3D区域传感器识别的零散工件中拾取零件，机器人手爪有较高的定位精度和抓持稳定性，定位精度高、耐用性好、维护简单；

（4）2D 视觉系统：该视觉系统由一个 2D 摄像头完成视觉数据采集。该视觉系统作为待加工工件准确抓取的定位方式，省去通常为满足机器人的准确抓取而必需的机械预定位夹具，具有很高的柔性；

（5）3D 散堆识别：该视觉系统由一个安装于地面上的3D广域传感器测出工件位置后，由机器人进行散堆拾取；

（6）上下料接驳台：实现物料周转箱的周转定位存放，便于立体料库的货叉上下料；

（7）立体料库：用于存放待加工及加工完成后工件，配备 WMS 管理软件，学生可以学习掌握立体料库出库的控制；

（8）输送料道：实现待加

工和加工完成后存放工件的周转箱的输送；

（9）数控车床：实现火箭部件及法兰盘的工序加工；

（10）比对仪：用于加工工件过程工序的检测，学生可以通过此检测台学习现代工厂最先进的检测技术；

（11）翻转机构：实现零部件加工姿态的翻转，结构简单，便于教学；

（12）固定机器人双手爪：用于机床上下料等工件夹持，具有较高的定位精度和抓持稳定性，定位精度高、耐用性好、维护简单；

（13）物料缓存台：实现待加工零部件的缓存存放或二次定位；

（14）立式加工中心：实现火箭模型部件及法兰盘的工序加工；

（15）天轨：实现两个立式加工中心之间物料的自动化上下料；

（16）立式加工中心夹具：自动化夹具，用于加工时将工件准确定位和可靠夹紧，满足加工需求和保证加工精度；

（17）清洁装置：实现对工件的自动吹气清洁；

（18）三坐标测量机：用于检测加工完成工件，学生可以

of the turnover box that stores the workpiece after processing and processing is completed;

(9)CNC lathe: to realize the processing of rocket parts and flanges;

(10)Comparator: used for the inspection of the work-piece process. Students can learn the most advanced testing technology in modern factories through this testing platform;

(11)Overturn mechanism: to realize the overturning of the processing posture of parts, simple structure, easy to teach;

(12) Fixed robot two-handed gripper: used for clamping workpieces such as loading and unloading of lathe, with high positioning accuracy and holding stability, high positioning accuracy, well durability, and simple maintenance;

(13)Material cache table: realize cache storage or secondary positioning of parts to be processed;

(14) Vertical machining center: to realize the processing of rocket model parts and flanges;

(15)Overhead rail: realize the automatic loading and unloading of materials between two vertical machining centers;

(16)Vertical machining center fixture: automatic fixture, used for accurate positioning and reliable clamping of the workpiece during processing, to meet processing requirements and ensure processing accuracy;

(17)Purging device: realize automatic blowing and cleaning of the workpiece;

(18)Three-coordinate measuring machine: used to detect the finished workpiece. Students can learn the most advanced testing technology in modern factories through this testing platform;

(19) Laser engraving machine: used to mark custo-mized pattern on the surface of the workpiece after processing;

(20)Assembly unit: realize the automated assembly of rocket crafts;

(21) Main control system: Use the man-machine interface to monitor the running status of the entire system, and use the PLC controller to realize the transmission of real-time and non-real-time data in the system, which is with high reliability and maintainability. The safety equipment adopts a gate switch as the safety protection of the robot's working area, which is completely isolated from man and machine to ensure the safety of personnel in the automatic operation of the system;

(22)Monitoring console: the operation of infor-mation systems such as MES system to control production and processing;

(23) Production line monitoring system: It can monitor and count the processing conditions, lathe status, movements and data of the machining center machine tools in real time;

(24)Safety fences and safety gates: Safety fences are mainly used in robot automation, and the safety protection of workshop equipment and logistics storage, ensuring the regional safety of personnel, production processes, products and property. It adopts a plexiglass panel and column structure as a whole,and its design meets mechanical safety standards. The safety door lock design is suitable for simple and instant installation on the safety door frame.

通过此检测台学习现代工厂最先进的检测技术；

（19）激光打标机：用于加工完成后工件表面进行定制化图案打刻；

（20）装配单元：实现火箭工艺品的自动化装配；

（21）主控系统：运用人机界面对整个系统的运行状态进行监控，采用 PLC 控制器，实现系统中实时和非实时数据的传输，具有高度可靠性和可维护性，安全设备采用门开关，作为机器人工作区域的安全防护，完全做到人机隔离，确保系统在自动运行中的人员安全；

（22）监控操作台：MES 系统等信息化系统的操作，实现对生产加工控制；

（23）生产线监控系统：能实时对加工中心机床的工件加工情况、机床状态、动作和数据进行监控、统计等；

（24）安全围栏及安全门：安全护栏主要应用于机器人自动化，以及车间设备和物流仓储的安全防护领域，确保人员、生产过程以及产品和财产的区域安全，它整体采用面板为有机玻璃及立柱结构，设计符合机械安全标准，安全门锁设计适用于在安全门框上进行简易、即时安装。

3. 智能制造综合生产线的加工流程

（1）火箭装配单元的物料周转流程：

①MES 系统下发指令，3DA 机器视觉对物料进行分拣，自动立体仓库出料；

②经传送带运送到数控车床，完成毛坯料的端面加工和外圆加工（一级火箭、二级火箭和助推器）；

③加工完成后，比对仪对其进行检测，合格后传送到立式加工中心，天轨机器人将半成品放在机床内进行加工，当工件完成最后一道加工工序后，机器人将成品放置在输送线上；

④检测单元的固定机器人将成品件放在三坐标检测仪检测台上，对 4 个助推器三级火箭外圆和孔的尺寸进行检测；

⑤检测单元的固定机器人将检测合格的成品放在打标机上，对二级火箭进行激光打标，后送至组装台完成组装；

⑥组装完成后，通过 AGV 在成品收集区将加工完成零件运输至自动立体仓库。

（2）法兰盘的加工流程：

①工件从立体料库出库，工业机器人通过 2D 工业相机

3.the processing flow of the intelligent manufacturing integrated production line

(1)material turnover process of the rocket assembly unit:

①MES system issues instructions, then 3DA machine vision sorts materials, and discharges materials from the automatic stereoscopic warehouse;

②It is transported to the CNC lathe through the conveyor belt to complete the end surface processing and outer circle processing of the blank material (first-stage rocket, second-stage rocket and booster);

③After processing, the comparator will test it and send it to the vertical machining center after passing the test. The overhead-rail robot will place the semi-finished product in the machine tool for processing. When the workpiece has completed the last procedure, the robot will place the finished product on the conveyor;

④The fixed robot of the inspection unit places the finished parts on the inspection table of the three-coordinate inspection instrument, and inspects the dimensions of the outer circle and the hole of the 4 boosters and the third-stage rocket;

⑤The fixed robot of the inspection unit puts the qualified products on the marking machine that performs laser engraving on the secondary rocket, and then sends it to the assembly station to complete the assembly;

⑥After the assembly is completed, the processed parts in the finished product collection area are transported to the automatic stereoscopic warehouse through the AGV.

(2) the processing flow of the flange:

①The workpiece is discharged from the stereoscopic warehouse, then the industrial robot recognizes the position of the workpiece through the 2D industrial camera, and places the workpiece on the conveyor line;

②The fixed robot of the CNC lathe processing unit grabs the workpiece and places the workpiece on the No. 1 CNC lathe for processing;

③After the processing is completed, the industrial robot takes out the workpiece and reverses it, and sends the workpiece to the No.2 CNC lathe for processing;

④After the processing is completed, the industrial robot puts the workpiece on the unload convey rack, then the guide rail takes the material to the vertical machining center unit for processing, and the overhead-rail robot of the vertical machining center grabs the workpiece and puts the workpiece to No.1 processing center for processing;

⑤After the processing of the No.1 machining center is completed, the overhead-rail robot takes out the workpiece and puts it into the No.2 vertical machining center for processing;

⑥After the processing of the processing unit of the machining center is completed, the overhead robot puts the workpiece on the unload convey rack, and the guide rail transfers the workpiece to the inspection unit;

⑦The fixed robot of the inspection unit puts the delivered workpiece on the three-coordinate inspection table for inspection;

⑧After the inspection is qualified, the fixed robot of the inspection unit puts the workpiece on the AGV, and the AGV transports the workpiece to the stereoscopic material warehouse for storage.

识别工件位置，并将工件放在输送线上；

②数控车床加工单元的固定机器人抓取工件，并将工件放在 1 号数控车床进行加工；

③加工完成后，工业机器人把工件取出并进行反转，把工件送入 2 号数控车床进行加工；

④加工完成后，工业机器人将工件放在出料接驳架上，导轨把物料带到立式加工中心单元进行加工，立式加工中心的天轨机器人抓取工件，并把工件放入 1 号加工中心进行加工；

⑤1 号加工中心加工完成后，天轨机器人把工件取出，并放入 2 号立式加工中心进行加工；

⑥加工中心加工单元加工完成后，天轨机器人把工件放在出料接驳架上，导轨把工件传送到检测单元；

⑦检测单元的固定机器人将送来的工件放在三坐标检测台上进行检测；

⑧检测合格后，检测单元的固定机器人把工件放在AGV上，并由 AGV 运送到立体料库入库。

Lecture 14-1
Introduction to Robot Studio(1)

微信扫一扫

教学视频

14-1 介绍机器人工作室（1）

1. 陕西国防工业职业技术学院介绍

陕西国防工业职业技术学院 1958 年建校，是一所由陕西省人民政府举办的全日制普通高等学校。学院先后隶属中国第三、第五机械工业部，中国兵器工业部，中国兵器工业总公司，是国家国防科工委重点支持院校。1999 年学院划转陕西省管理，2001 年升格为全日制普通高等职业院校，是中国兵器工业集团、国家国防科工局与陕西省“省部共建”院校，陕西国防工业职教集团牵头单位，中国航天科技教育联盟副理事长单位。建校以来，学院立足陕西，面向国防，为我国国防科技工业和经济社会发展培养了大量创新型高素质技术技能人才，学子遍布全国十二大军工集团。图 14-1 为该校校园风景。

1. Introduction to Shaanxi Institute of Technology

Shaanxi Institute of Technology was established in 1958. It is a full-time general college sponsored by People's Government of Shaanxi Province. The college is successively affiliated to the third and fifth Ministry of Machine-Building Industry of the PRC, the Ministry of Ordnance Industry of PRC, and the NORINCO(G). It is a key college supported by the COSTIND. In 1999, the college was transferred to Shaanxi Province for management. In 2001, it was upgraded to a full-time general higher vocational college. It is a college "jointly established by China North Industries Group, State Administration of Science. Technology and Industry for National Defense and Shaanxi Province", the leading unit of Shaanxi National Defense Industry Vocational Education Group, and the vice chairman unit of China Aerospace Science and Technology Education Alliance. Since its establishment, the college, based in Shaanxi province and oriented to national defense, has trained a large number of innovative and high-quality talents with technical skills for the development of national defense science and technology industry, economy and society, and students have spread throughout the country's 12 major military industry groups. The campus scenery of the college is shown in Fig. 14-1 below.

Fig.14-1 The campus scenery of our college
图 14-1 校园风景

2. 工业机器人应用人才培养中心建设背景与目标

工业机器人应用人才培养中心项目来源于教育部与3家公司（华航唯实、ABB、新时达，统称实施方）在2016年开展的工业机器人领域职业教育合作项目。该合作项目旨在推动国务院《中国制造2025》规划的实施，进一步推进制造强国战略，为我国的工业机器人领域提供急需的高素质技能技术型人才。

作为该项目在陕西省的首家建设启动单位，该校的工业机器人应用人才培养中心除了作为我校的工业机器人技术专业教学基地外，更担负着推动地方企业单位和相关院校的技术水平和科研及技术服务能力提升的重任。因此，提出了以培养工业机器人专业技术技能人才，建成国内领先、陕西一流的技术技能人才培养中心和职业技能鉴定基地，为区域内社会企业提供机器人技术开发和科研任务作为我校工业机器人应用人才培养中心的建设目标。

3. 工业机器人应用人才培养中心建设思路

通过与华航唯实等实施方合作，实现校企联动办学、联

2. Construction Background and Goals of the Industrial Robot Application Talent Training Center

The Industrial Robot Application Talent Training Center project originated from the vocational education cooperation project in the field of industrial robots carried out by the Ministry of Education and 3 companies (Huahang Weishi, ABB, and Xinshida, collectively referred to as the implementers) in 2016. The purpose of this cooperation project is to promote the implementation of the "Made in China 2025" plan of the State Council , further advance the strategy of building China into a manufacturing power, and provide urgently needed high-quality skilled technical personnel for China's industrial robot sector.

As the first construction start-up unit of the project in Shaanxi Province, the Industrial Robot Application Talent Training Center of the college not only serves as the teaching base for industrial robot technology major of our college, but also responsible for promoting the technical level, scientific research and technical service capabilities of local enterprises and relevant colleges and universities. Therefore, it is proposed to cultivate industrial robots professional and technical talents, build a domestic leading and Shaanxi first-class technical talent training center and vocational skills appraisal base, and provide robot technology development and scientific research for enterprises in the region as the construction goal of the Industrial robot application talent training center.

3. Construction Ideas of Industrial Robot Application Talent Training Center

Through the cooperation with Huahang Weishi

and other implementing parties, we can realize school-enterprise linkage education, linkage training, linkage research and development, and achieve the organic integration of curriculum system and professional positions, course content and job capabilities, professional teachers and skilled craftsmen, training bases and production workshops, school evaluation and social evaluation, campus culture and corporate culture. The logo of Huahang Weishi is shown in Fig. 14-2.

动培训、联动研发，达到课程体系与专业岗位、课程内容与岗位能力、专业教师与能工巧匠、实训基地与生产车间、学校评价与社会评价、校园文化与企业文化的有机融合。图14-2为华航唯实企业标识。

Fig. 14-2 Huahang Weishi
图14-2 华航唯实

It can provide colleges with precise corporate employment requirements, jointly set up orientation classes and order classes, and promote high-quality students to practice in the company. At the same time, the urgently needed pre-employment training and skill improvement training needs of employees will be brought back to the college to realize the hematopoietic function of the training base and increase the utilization rate of equipment.

公司可以为学校提供精确的企业用工需求，共同开设定向班、订单班，促成优质学生到企业中实习锻炼。同时将企业急需的职工岗前培训、技能提升锻炼需求引回学校，实现实训基地的造血功能，提高设备使用率。

At the same time, the employment requirements of many well-known domestic and foreign industrial robot manufacturers can be released in the employment section of the online platform, and the employment guidance services for qualified graduates of the college can be provided by employment guidance recommendation module. The module organically integrates corporate information and

同时可将多家国内外工业机器人知名厂商的用工岗位需求在网络平台就业板块中发布，通过就业指导推荐模块为学校的合格毕业生提供就业指导服务。模块将企业信息进行有机整合，结合大数据分析技术，对学校即将就业学生进行针对性推荐，提高学生就业率，满足企业的招工需求，彻底解决人才出口问题。对于学生来说，就业推荐功能模块提供了所有企业所有岗位技能要求，使其在学习前就明确目标，对部分知识点有针对性地重点学习，有效缩短岗位适应期，提高就业竞争力。对于企业用户来说，模块可以通过大数据分析和匹配度计算，快速寻找企业所需技能人才，提高招聘效率，实现企业和学校的双赢。

4. 工业机器人应用人才培养模式

工业机器人应用人才培养中心是学校的工业机器人技术专业建设的载体和依托，该中心依然以培养学生为第一要务，按照“整体规划，分步实施，够用为度，适度超前”的建设理念，本着学生乐学、教师易教的原则，充分利用多方资源，校企共建，融入新形态教育形式和“互联网+”理念，充分发挥现代化教学资源和实施方所提供的网络平台优势的建设思路，同时依据“虚拟、认知、基础、应用、综合”的功能分区布局，形成“理论学习→仿真训练→岗位实操”的“虚实结合，能力递进”的人才培养模式。如图 14-3 所示。

combines big data analysis technology to make targeted recommendations to the students who are going to work of our college, which will increase the employment rate of students, meet the recruitment needs of enterprises, and thoroughly solve the problem of talent export. For students, the employment recommendation function module provides all the job skills requirements of all companies, so that they can clarify their goals before learning, and focus on certain knowledge points, effectively shorten the job adaptation period and improve employment competitiveness. For enterprise users, skilled talents needed by enterprises can be quickly found on the module through big data analysis and matching calculation, which can improve recruitment efficiency, and achieve a win-win cooperation between enterprises and colleges.

4.Industrial Robot Application Talent Training Model

The construction of the Industrial Robot Application Talent Training Center is the carrier and support for the construction of the industrial robot technology major of the college. The center still takes the training of students as the first priority. In accordance with the construction concept of “overall planning, step-by-step implementation, adequate use, and moderate advancement”, in line with the principles of students’ enjoyment of learning and teachers’ easy teaching, we make full use of multiple resources, jointly build by colleges and enterprises, and integrate new forms of education and the Internet+ concept. We give full play to the construction idea of the advantages of modern teaching resources and the network platform provided by the implementer, at the same time, according to the "virtual, cognitive, basic, application, and comprehensive" function division layout, form the "theoretical learning → simulation training →post practice" talent training model of "combination of virtual and reality, progressive ability". As shown in Fig. 14-3 below.

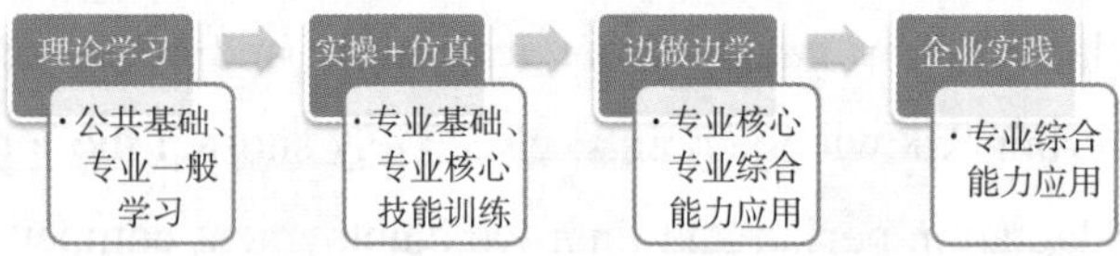

Fig. 14-3 Talent Training Model
图 14-3 人才培养模式

Theoretical learning: refers to the study of related theoretical courses in the professional basic learning area, the professional general learning area, the professional core learning area and the professional development learning area. At this stage, the teaching model of "teaching, learning and practising" is always running through, following the students' cognitive rules, and arranging the content of the course from the shallower to the deeper, so that the students can "learn by practising", so as to realize the progressive improvement and promotion of knowledge and ability.

Simulation training: In view of the specific characteristics of the industrial robot application industry, introduce industrial robot offline programming simulation software into the teaching process of "Industrial Robot Technology Basics", "Industrial Robot Offline Programming Training" and other professional core courses, make full use of modern virtual education technology, establish virtual teaching

理论学习：指专业基础学习领域、专业一般学习领域、专业核心学习领域及专业拓展学习领域相关理论课程的学习。在此阶段，始终贯穿“教、学、做”于一体的教学模式，遵循学生认知规律，由浅入深地安排课程内容，使学生在“做中学”，从而实现知识及能力的逐级递进和提升。

仿真训练：考虑到工业机器人应用行业的具体特点，在《工业机器人技术基础》《工业机器人离线编程实训》等专业核心课程教学过程中引入工业机器人离线编程仿真软件，充分运用现代虚拟教育技术，建立仿真项目等虚拟教学环境，集中在校内交替进行具体产品生产工艺、操作过程的虚拟仿真训练。校内专业教师与企业能工巧匠共同担任指导教师，训练内容涉及工艺流程、操作原理、设备原理、控制原理、事故现象、产生原因和排除方法等多方面知识。在仿真训练过程中，学生利用计算机操作设备，调节操作参数，模拟真实的工业机器人操作，从而掌握现场操作，为后续学习打下良好的基础。

岗位实操：在理论学习、技能训练及仿真训练的基础上，在工业机器人应用人才培养中心和合作企业进行生产性实训及顶岗实习，进行和企业产品生产相适应的专业核心技能学习。顶岗实习时，企业的能工巧匠担任指导教师，学生在实习基地以职业人的身份参与企业生产活动，承担工作岗位规定的责任和义务，增加了学生对生产过程——包括工业原理、工艺流程、生产设备、规章制度等的切身认识，使学生及时掌握最新工艺和技能，强化学生的专业能力、协作精神和责任意识，使学生的课堂知识真正转化成工作能力，逐渐培养学生向职业人转变，实现人才培养与企业需求的最大程度对接。

在“理论学习→仿真训

environments such as simulation projects, and focus on alternate virtual simulation training of specific product production processes and operating processes in the college. The professional teachers in the college and the skilled craftsmen of the enterprise jointly act as instructors. The training content involves many aspects of knowledge such as process flow, operation principle, equipment principle, control principle, accident phenomenon, cause and elimination method. In the simulation training process,students use computers to operate equipment, adjust operating parameters, and simulate real industrial robot operations, so as to master on-site operations and lay a good foundation for subsequent learning.

Post practice: On the basis of theoretical study, skill training and simulation training, carry out productive training and on-the-job internships in the Industrial Robot Application Talent Training Center and partner companies, and conduct professional core skills learning that are compatible with the production of the company's products. During the internship, the skilled craftsmen of the enterprise serve as instructors. The students participate in the production activities of the enterprise as professionals in the internship base, and assume the responsibilities and obligations stipulated by the job position. This increases the students' personal knowledge of the production process-including industrial principles, technological processes, production equipment, rules and regulations, etc., enables students to master the latest technology and skills in a timely manner, strengthens students' professional abilities, collaboration spirit and sense of responsibility, so that students' knowledge learned in classroom is truly transformed into work abilities, and students are gradually trained to transform into professionals, which achieves the maximum connection between talent training and enterprise needs.

In the "theoretical learning→simulation training →post practice" ability advancement process, based on job skills and professional quality requirements, the quality education module is infiltrated into the professional teaching system. Respect the subjectivity and initiative of students, pay attention to the development of students' intellectual potential and personality, and promote each student to get a comprehensive and active health education. In teaching activities, ideological and moral education is the core to cultivate students' noble character and excellent quality; based on humanistic quality education, improve students' humanistic accomplishment and cultural taste; focus on professional quality education to enhance students' core competitiveness; take physical and mental quality education as a guarantee, cultivate students' good physical fitness and mentality, and integrate quality education throughout students' learning; strengthen practical teaching, implement the "Dual Certificate" System through the ability advancement of "theory learning → simulation training → post practice", and cultivate the first-line technical and technical professionals in production, construction, service and management.

练→岗位实操”能力进阶过程中，基于岗位技能及职业素养需求，将素质教育模块渗透到专业教学体系中。尊重学生的主体性和主动性，注重开发学生的智慧潜能和个性，促进每个学生全面地、主动地得到健康的教育。在教学活动中，以思想道德教育为核心，培育学生的高尚品格和优秀品质；以人文素质教育为基础，提高学生的人文修养和文化品位；以职业素质教育为重点，提升学生的核心竞争力；以身心素质教育为保障，培养学生良好的身体素质和心态，将素质教育贯穿学生学习始终；加强实践教学，通过“理论学习→仿真训练→岗位实操”能力进阶，实行“双证书”制，培养生产、建设、服务、管理第一线的技术技能型专门人才。

微信扫一扫

教学视频

Lecture 14-2
Introduction to Robot Studio(2)
14–2 介绍机器人工作室（2）

1. 工业机器人应用人才培养中心总体规划

在调研和专家论证的基础上，统一规划，合理布局，最终建成如下的工业机器人专业实训条件，如图 14–4、14–5 所示。工业机器人应用人才培养中心按照循序渐进，由虚拟到真实、由编程到应用，逐步提高的教学步骤设计建设，最终形成一个完整地支撑专业发展的实训网络。

1. Overall Plan of the Industrial Robot Application Talent Training Center

On the basis of research and expert argumentation, with unified planning and reasonable layout, the following professional training conditions for industrial robots have been finally established, as shown in Fig. 14-4 and 14-5. The Industrial Robot Application Talent Training Center is designed and constructed step by step, from virtual to real, from programming to application, gradually improved teaching techniques, and finally forms a complete training network to support professional development.

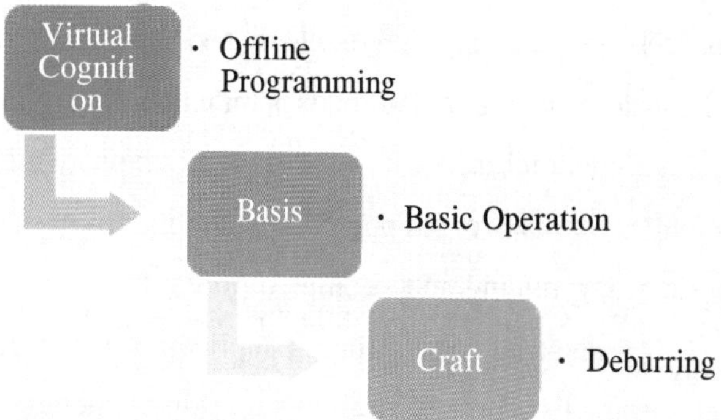

Figure.14-4 Professional Training Conditions
图 14–4 专业实训条件

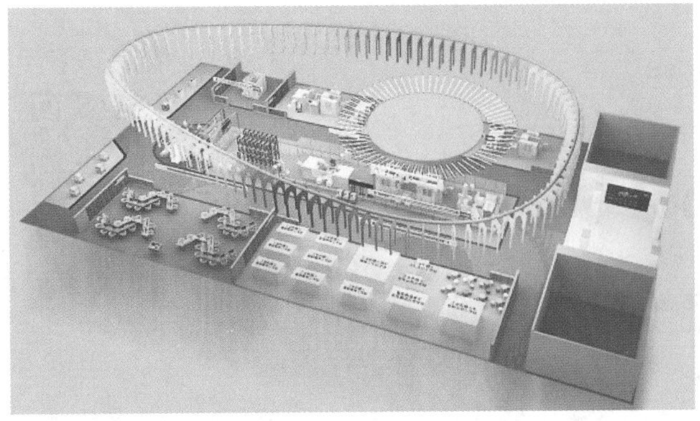

Figure. 14-5 Construction Plan
图 14-5 建设规划图

(1)Virtual industrial robot teaching: integrate the teaching characteristics of theory, virtual and real, carry out simulation teaching of industrial robot courses, and conduct offline programming training and application for students.

(2)Cognitive industrial robot teaching: theoretical teaching followed by ontology cognition, virtual simulation followed by ontology operation, to help students understand the operating principles of the mechanical transmission system and control system of industrial robot.

(3) Industrial robot basic teaching: It not only focuses on teaching the function and operation of industrial robots, but also focuses on teaching applied technology combined with process realization, which provides hardware support for students to master basic programming and simulated process.

(4)Industrial robot process application teaching: relying on two process training platforms including deburring and intelligent manufacturing unit system integration application platform, students will be trained on industrial robots combined process.

（1）虚拟工业机器人教学：综合理、虚、实一体化教学特征，开展工业机器人课程仿真教学，面向学生进行离线编程实训和应用。

（2）认知工业机器人教学：先理论教学后本体认知，先虚拟仿真后本体操作，帮助学生理解工业机器人机械传动系统和控制系统的工作原理。

（3）工业机器人基础教学：既针对工业机器人本身的功能和操作教学，又着眼于结合工艺实现的应用技术教学，为学生掌握基础编程和模拟工艺能力操作提供硬件支撑。

（4）工业机器人工艺应用教学：依托去毛刺、智能制造单元系统集成应用平台等两个工艺实训平台，对学生进行结合工艺的工业机器人实训。

2. 工业机器人应用人才培养中心工作站

2.Industrial Robot Application Talent Training Center Workstation

Fig.14-6 Industrial Robot Ontology Cognitive Workstation

图14-6 工业机器人本体认知工作站

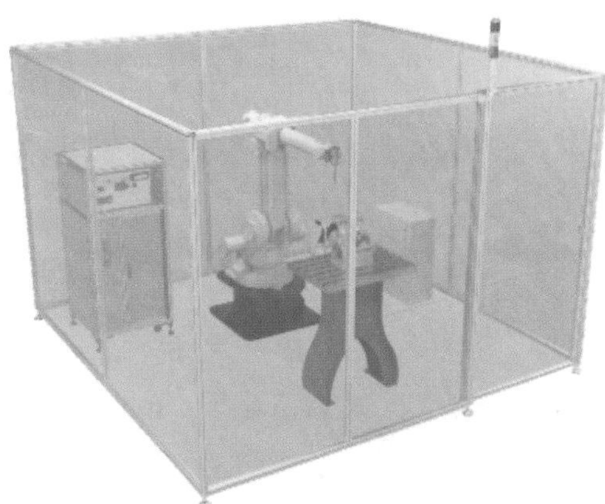

Fig.14-7 Industrial Robot Deburring Workstation

图14-7 工业机器人去毛刺工作站

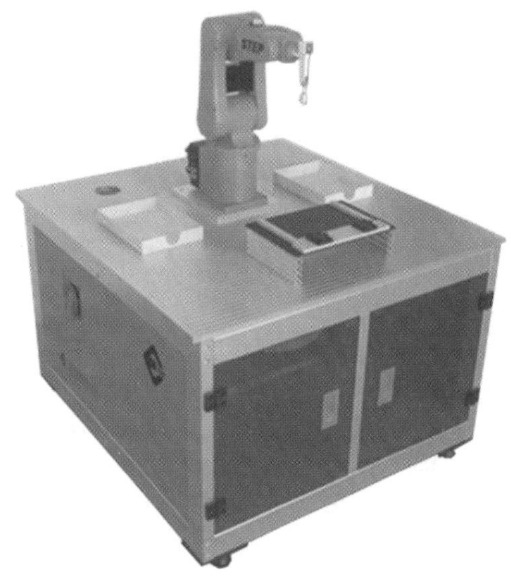

Fig.14-8 Industrial Robot Human-machine Interactive Workstation

图14-8 工业机器人人机互动工作站

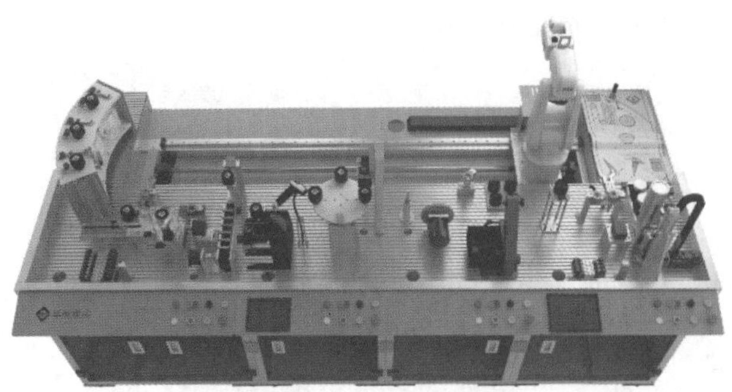

Fig.14-9 Industrial Robot Basic Teaching Workstation-Multi-Process

图 14-9 工业机器人基础教学工作站-多工艺

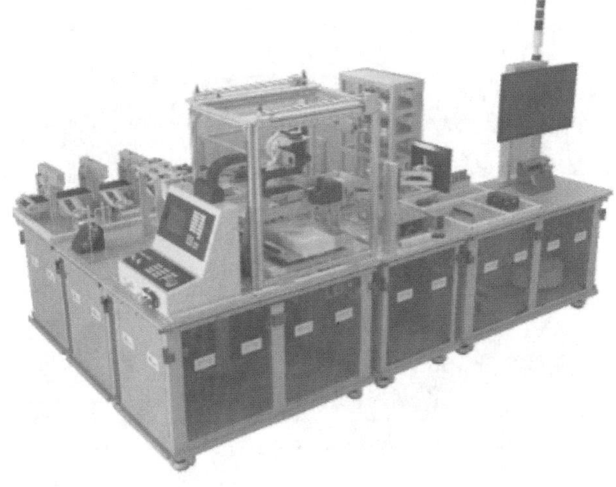

Figure.14-10 Intelligent Manufacturing Unit System Integration Application Platform

图 14-10 智能制造单元系统集成应用平台

Figure.14-11 Industrial Robot Basic Teaching Workstation-Maintenance

图 14-11 工业机器人基础教学工作站-维护维修

（1）工业机器人本体认知工作站

工业机器人本体认知工作站作为学生对工业机器人核心部件和机械结构的认知平台，主要应用在工业机器人技术教学的初期。通过从内部机械构成来理解工业机器人的运动方式和动力传递，有助于学生更简易地消化理论知识，完善机械设计及运动控制的知识结构。该工作站通过对其进行零件级的解构，按照真实装配结构顺序进行排列组合，将主要核心部件固定在展架上，配合灯光及背景图进行展示。为保证教学的真实性，所有零件均来自真实应用的工业机器人。

（2）工业机器人去毛刺实训工作站

工业机器人去毛刺实训工作站主要面向具有一定的工业机器人基础操作能力的学生，以真实工厂应用要求为基础，提供结合去毛刺工艺的工业机器人实训平台。该实训工作站由工业机器人、去毛刺工具、供气系统、工作台及工装夹具、工业吸尘器、安全防护组件及配套设备构成，利用工业机器人这一柔性执行单元，结合去毛刺工艺实现的实际加工单元，可以完成对任意形状零件

(1)Industrial Robot Ontology Cognitive Workstation

Industrial robot ontology cognitive workstation, as a platform for students to recognize the core components and mechanical structure of industrial robots, is mainly used in the early stage of industrial robot technology teaching. Understanding the motion mode and power transmission of industrial robots from the internal mechanical structure will help students digest theoretical knowledge more easily and improve the knowledge structure of mechanical design and motion control. The workstation deconstructs robots at the part level, arranges and combines them in the order of the actual assembly structure, fixes the main core components on the display rack, and displays them with lighting and background pictures. In order to ensure the authenticity of teaching, all parts are from industrial robots in real applications.

(2)Industrial Robot Deburring Training Workstation

The industrial robot deburring training workstation is mainly for students who have certain basic operating capabilities of industrial robots, which provides an industrial robot training platform that combines deburring technology based on the requirements of real factory applications. The training workstation is composed of industrial robots, deburring tools, air supply systems, workbenches and fixtures, industrial vacuum cleaners, safety protection components and supporting equipment. Using the flexible execution unit of the industrial robot, combined with the actual processing unit realized by the deburring process, the deburring process of any shape parts can be completed.

(3) Industrial Robot Human-machine Interactive Workstation

The industrial robot human-machine interactive workstation, with industrial robots as the core, combined with wireless communication, image processing, and offline programming technology, enables people to write and draw freely on the tablet. Industrial robots can reproduce handwriting on the drawing board, which reflects the real-time control application t echnology of industrial robot. Meanwhile, people can fully participate in the operation of the workstation, which enhance the interest of teaching and the display of equipment.

(4)Industrial Robot Basic Teaching Workstation-Multi-Process

The multi-process workstation for basic teaching of industrial robots is mainly applied in the basic operation teaching stage of industrial robots, aiming at the basic parameter setting, trajectory progra-mming of industrial robots, as well as the training teaching combined peripheral application equipment and simulated processing technology.With the industrial robot as a platform, combined with comprehensive technologies such as automatic control, processing technology, visual inspection, tag recognition, a ssembly inspection, etc., the workstation are the best hardware platform for industrial robot technology teaching, which helps students understand the application of industrial robots from the system level, and improves the basic technological operating ability and debugging skills to meet multi-level personalized teaching needs.

的去毛刺加工。

（3）工业机器人人机互动工作站

工业机器人人机互动工作站，以工业机器人为核心，结合无线通信、图像处理、离线编程技术，实现了人在平板上随意写字绘画，工业机器人可在画板上复现笔迹，体现了工业机器人的实时控制应用技术，同时让人充分地参与到工作站的运行过程中，增强教学的趣味性和设备的展示性。

（4）工业机器人基础教学工作站-多工艺

工业机器人基础教学多工艺工作站主要应用在工业机器人基础操作教学阶段，针对工业机器人基本参数设置、轨迹编程以及结合周边应用设备和模拟加工工艺的实训教学，该工作站以工业机器人为平台，融合了自动控制、加工工艺、视觉检测、标签识别、装配检测等综合技术，是进行工业机器人技术教学的最佳硬件平台，有助于学生从系统级了解工业机器人的应用方式的操作方法，完善基础技术操作能力和调试技能，满足多层次个性化教学需求。

（5）工业机器人智能制造单元系统集成应用平台

全国高等职业院校技能大赛中“制造单元智能化改造与集成技术”赛项指定设备为北京华航唯实机器人科技股份有限公司的 CHL-DS-11 智能制造单元系统集成应用平台，该设备既能满足学生技能训练要求，又能满足参加职业技能大赛要求，可完成编程控制、I/O 通信、视觉系统、数控机床、坐标设定等项目。

(5) Industrial Robot Intelligent Manufacturing Unit System Integration Application Platform

The designated equipment for the "Intelligent Transformation and Integration Technology of Manufacturing Units" in the National Skills Competition for Higher Vocational Colleges is the CHL-DS-11 intelligent manufacturing unit system integration application platform of Beijing Huahang Weishi Robot Technology Co., Ltd. It can meet the requirements of students' skill training as well as the requirements of participating in vocational skill competitions. It can complete programming control, I/O communication, vision system, CNC machine tool, coordinate setting and other projects.

（6）工业机器人基础教学工作站-维护维修

工业机器人基础教学工作站-维护维修，以工业机器人本体及其控制系统最常见的维护维修操作为核心，学生通过对典型故障的诊断和排除，按照规范步骤完成对工业机器人本体的拆装过程，并根据图纸要求搭建控制系统并接线，由零件级别深入了解工业机器人的构成形式和运动方式，掌握基础维护维修技能。工作站配套了与生产厂家完全相同的工装、工具及测试设备，让学生可以在教学环境中充分体验高

(6) Industrial Robot Basic Teaching Workstation-Maintenance

Industrial robot basic teaching workstation-maintenance is centered on the most common maintenance of the industrial robot body and its control system. Through the diagnosis and elimination of typical faults, students complete the disassembly process of the industrial robot body according to the standardized steps, then build the control system and connect wires according to the requirements of the drawings, which gain an in-depth understanding of the composition and movement modes of industrial robots at the part level, and master basic maintenance skills. The workstation is equipped with the same frocks, tools and test equipment as the manufacturer, so that students can fully experience the production process requirements of high-precision electromechanical equipment in the teaching environment, and indeed master the maintenance and repair skills.

3.Construction of Teaching Resource Platform of the Industrial Robot Application Talent Training Center

Based on the Robot Application Talent Training Center, we build a professional teaching resource library for industrial robot technology with rich resources, advanced technology, open sharing, and sustainable development, making it a platform for teacher-student interaction to promote the reform of professional teaching methods and manners, become a lifelong learning digital training center for enterprise employees, and form a dynamic update mechanism for the resource library, as well as online resources of 5 courses including "Industrial Robot Technology Basics", "Industrial Robot Offline Programming", "Industrial Robot Technology and Application" and "Industrial Robot System Integration and Application", which has improved the quality of training professionals in industrial robot technology.

4. Social Training and Technical Services of the Industrial Robot Application Talent Training Center

Relying on the Industrial Robot Application Talent Training Center, Robot Innovation Studio and Resource Library Platform, they provide online learning s ervices such as resource retrieval, information inquiry, data download, teaching guidance for the continuing education of enterprise employees and social personnel, which solve the practical technological questions for enterprises. In the absence of a national professional qualification certificate for industrial robot technology,

精度机电设备的生产工艺要求，切实掌握维护维修技能。

3. 工业机器人应用人才培养中心教学资源平台建设

依托机器人应用人才培养中心，建成资源丰富、技术先进、开放共享、持续发展的工业机器人技术专业教学资源库，使其成为师生互动的平台，推动专业教学方法与手段改革，成为企业员工终身学习的数字化培训中心，形成资源库动态更新机制，建设“工业机器人技术基础”“工业机器人离线编程”“工业机器人工艺与应用”和“工业机器人系统集成与应用”等5门课程的在线资源建设，提升工业机器人技术专业人才培养质量。

4. 工业机器人应用人才培养中心社会培训与技术服务

依托工业机器人应用人才培养中心、机器人创新工作室和资源库平台为企业在职员工和社会人员的继续教育提供资源检索、信息查询、资料下载、教学指导等在线学习服务，为企业解决实际中的技术问题提供帮助，在目前没有国家工业机器人技术专业的专门职业资格证书的情况下，为提高入职的适应度，以工业机器人编程员和工业机器人操作维护员制定培训计划，开展社会培训工作，每年为相关企业员工培训100人次，并积极开展相关技能认证工作。为适应工业机器人技术的发展，拓展不同学习者的学习要求，通过校企合作的方式，选择国际国内行业领先单位的企业认证作为学习的资源之一。选择行业内国际国内知名企业 ABB、华航唯实、发那科（FANUC）等相应的认证，为校内外开展相关认证培训人数120人。图14-12所示为工业机器人应用人才培养中心建设图。

建成后的工业机器人应用人才培养中心将成为区域内高职院校技能交流的平台，积极申报国家级师资培训基地，为

in order to improve the adaptability of entry, we develop training plans and carry out social training work for industrial robot programmers and industrial robot operation and maintenance personnel. We have trained 100 employees for relevant enterprises and actively carried out relevant skills certification. In order to adapt to the development of industrial robot technology and expand the learning requirements of different learners, college-enterprise cooperation is selected as a way to get the certification. We have carried out relevant certification training with well-known international and domestic companies in the industry, such as ABB, Huahang Weishi, and Fanuc. Now there are 120 people who carry out relevant certification training inside and outside the college. Fig. 14-12 shows the location of the Industrial Robot Application Talent Training Center.

After construction, the Industrial Robot Application Talent Training Center will become a platform for skills exchanges between higher vocational colleges in the region, actively applying for national teacher

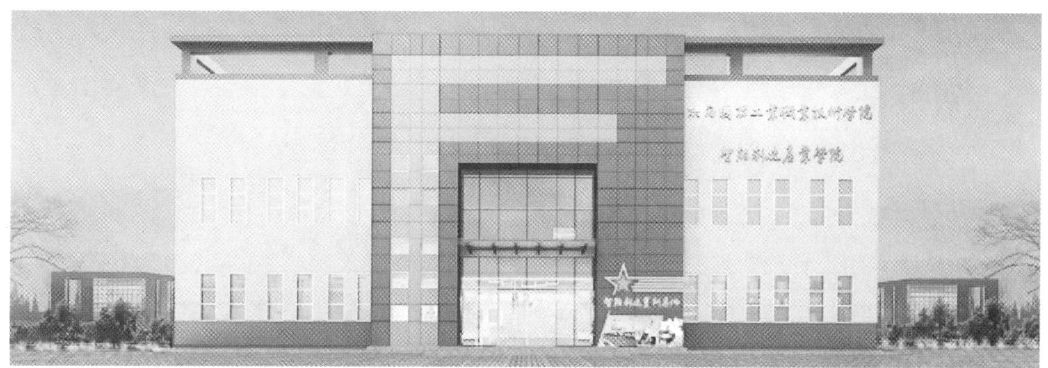

Fig.14-12 Construction of the Industrial Robot Application Talent Training Center

图14-12 工业机器人应用人才培养中心建设图

training bases, and providing teacher training services for higher vocational colleges across the country; we provide corresponding counterparts support for 3 secondary vocational colleges including Yulin Forestry School, Shaanxi Mechanical and Electrical Engineering School in terms of profession construction, teaching reforms, training conditions construction, and teachers' training and cultivating of industrial robots.

全国高职院校提供师资培训服务；为榆林林业学校、陕西机电工程学校等省内 3 所中职院校就工业机器人方向的专业建设、教学改革、实训条件建设和师资培训培养等方面开展对口支援工作。

Lecture 15-1
A Modeling for Industrial Handling Robots and and Simulation Workstation(1)

微信扫一扫

教学视频

15-1 工业搬运机器人的建模与仿真工作站（1）

RobotStudio6 软件可以做一些动画仿真效果，通过在软件里做动画仿真效果，我们可以节省机器人工作站的开发时间，提高效率，也可以提前发现工业机器人在工作站运行时遇到的一些问题。从这节课开始，我们开始学习 RobotStudio6 软件的动画仿真功能和如何在软件里做动画仿真效果。

RobotStudio6 软件的动画效果开发工具有两个，事件管理器和 SMART 组件。事件管理器简单易学，适合制作简单的、逻辑关系不多的动画仿真；SMART 组件适合做信号比较多、逻辑关系复杂的动画仿真，需要使用者系统学习工业机器人技术后才能使用。

接下来，在 RobotStudio6 软件里设计一个机械装置，制作步骤如下：

Some animation simulation effects can be done through RobotStudio6 software. By doing animation simulation in the software, we can save the development time of the robot workstation, improve efficiency, and discover some problems encountered by industrial robots in the workstation in advance. From this lesson, let's begin to learn the animation simulation function of RobotStudio6 software and how to make animation simulation effects in the software.

There are two animation effect development tools in RobotStudio6 software, event manager and SMART components. The event manager is simple and easy to learn, and is suitable for making simple animation simulations with few logical relationships; SMART components are suitable for animation simulations with more signals and complex logical relationships. Users need to learn industrial robot technology systematically before they can use it.

Next, we design a mechanical device in RobotStudio6 software, the production steps are as follows:

(1) Open the RobotStudio6 software, select [File]→[New]→[Empty Workstation], click [Create] to create an empty workstation, as shown in Fig. 15-1.

（1）打开 RobotStudio6 软件，选择【文件】→【新建】→【空工作站】，点击【创建】，创建一个空的工作站，如图 15-1 所示。

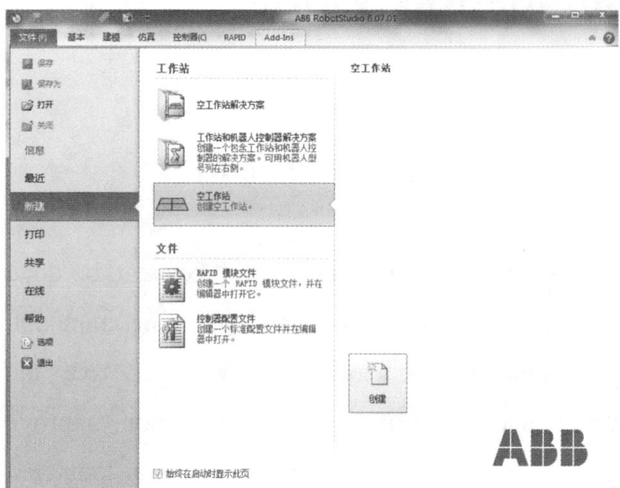

Fig. 15-1 Create an Empty Workstation

图 15-1 创建一个空的工作站

(2) In the [Basic] of menu bar, select [ABB Model Library]→ select [IRB120]→ click [OK].the operation process and effect are shown in Fig. 15-2.

（2）在菜单栏【基本】里，选择【ABB 模型库】→选择【IRB120】→点击【确定】。操作过程和结果如图 15-2 所示。

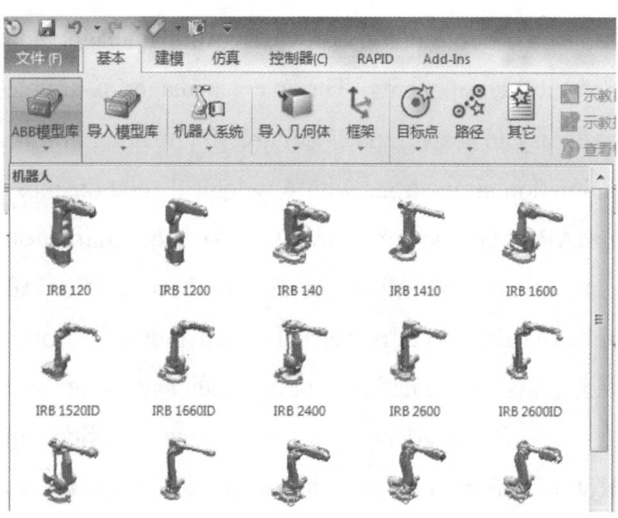

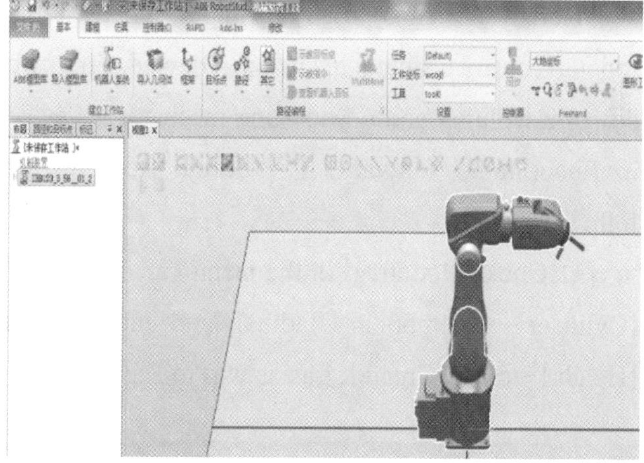

Fig.15-2 Select Industrial Robot
图 15-2 选择工业机器人

(3)点击【机器人系统】→【从布局】→设置系统名称和位置路径，位置路径必须是英文路径，并选择电脑上安装的 Robotware，点击下一个直到完成，如图 15-3 所示。

(3)Click [Robot System]→[From Layout...]→ Set the system name and location path, the location path must be an English path, and select Robotware installed on the computer, click Next until it is completed,,as shown in Fig. 15-3.

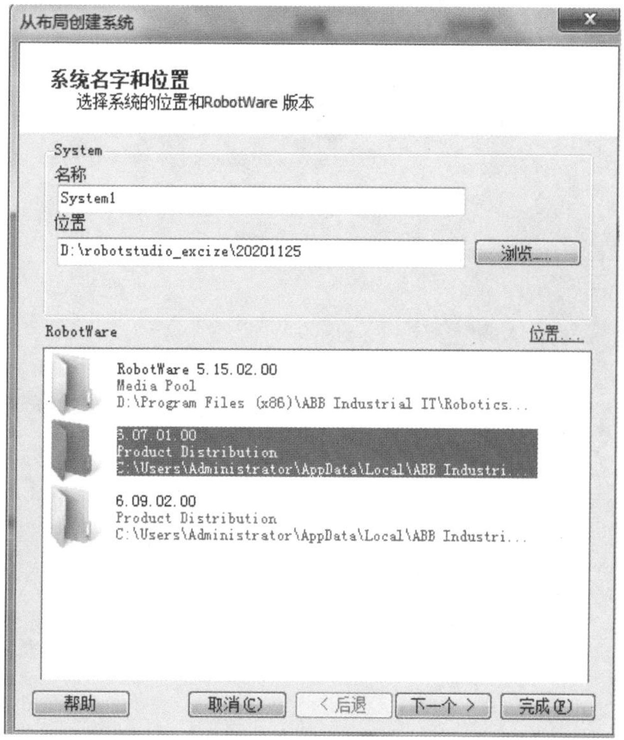

Fig 15-3 Create System from Layout
图 15-3 从布局创建系统

Here we need to use the output signal of the robot to control a simple mechanical device. First, we design a Column and a Slip Ring mechanical device in RobotStudio software. The design steps are as follows:

(1)Click [Modeling] in the menu bar→[Solid]→[Cylinder]→enter 50 in [Radius], and enter 600 in [Height]→click [Create],],as shown in Fig. 15-4.

这里我们要用机器人的输出信号去控制一个简单的机械装置，首先我们在 RobotStudio 软件里设计一个立柱和滑环的机械装置。设计步骤如下：

（1）点击菜单栏中的【建模】→【固体】→【圆柱体】→【半径】输入 50，【高度】输入 600→点击【创建】，如图 15-4 所示。

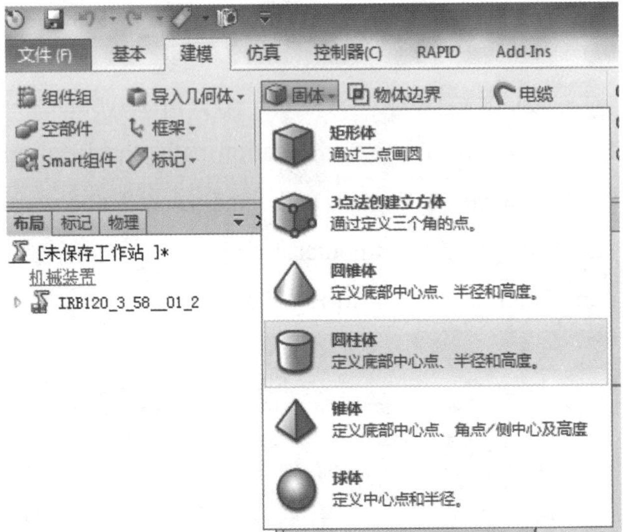

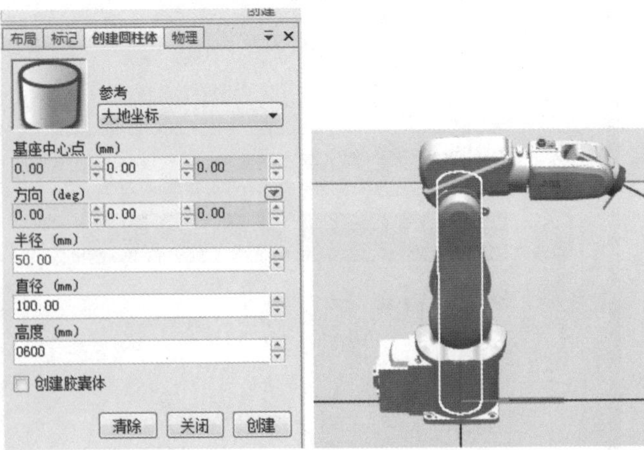

Fig.15-4 Create Column

图 15-4 创建立柱

（2）制作一个滑环，【半径】输入 100，【高度】输入 50→点击【创建】，如图 15-5 所示。

(2) Making a Slip Ring, enter 100 in [Radius], enter 50 in [Height] → click [Create], as shown in Fig. 15-5.

Figure.15-5　Create Slip Ring

图 15-5　创建滑环

（3）现在我们可以看到左边的【布局】里面，部件_1 是立柱，部件_2 是滑环。现在机器人和部件_1 以及部件_2 是重叠的，所以需要把机器人隐藏起来，在【IRB120】上面点击鼠标右键，把【可见】前面的对钩去掉，机器人消失，如图 15-6 所示。

(3) Now we can see in [Layout] on the left, component_1 is a Column and component_2 is a Slip Ring. Now the robot is overlapped with component_1 and component_2, so you need to hide the robot. Right-click [IRB120], remove the check mark before [Visible], and the robot disappears, as shown in Fig. 15-6.

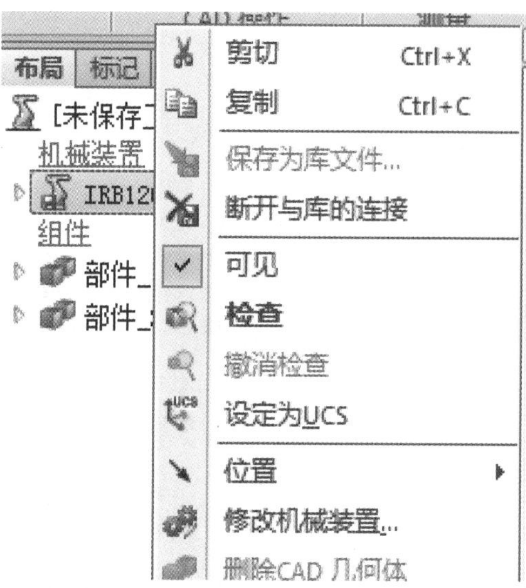

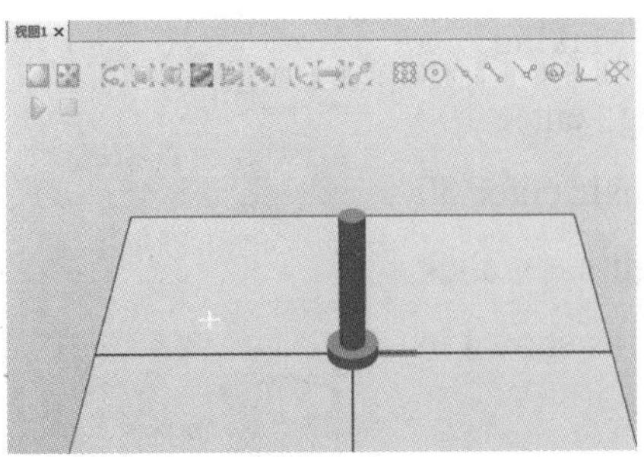

Fig.15-6 Hide the Robot

图 15-6 隐藏机器人

(4) Right-click component_2 in [Layout], select [Rename] → enter “Slip Ring”, and similarly rename component_1 to “Column”, as shown in Fig. 15-7.

（4）在【布局】里在部件_2上点击鼠标右键，选择【重命名】→输入“滑环”，同理把部件_1 重命名为“立柱”，如图 15-7 所示。

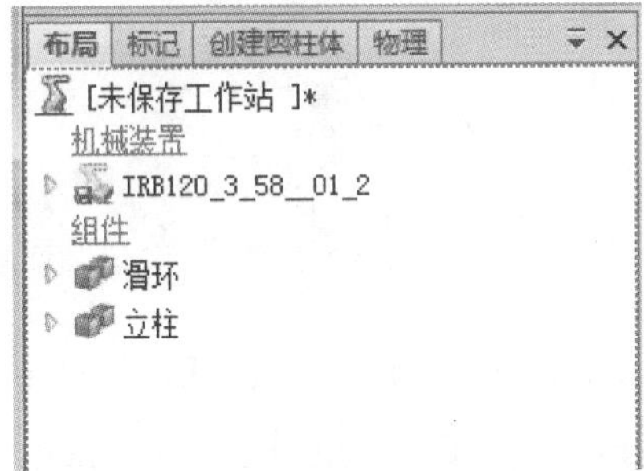

Fig.15-7 Modify Component Name

图 15-7 修改部件名称

(5) Right-click the Slip Ring in [Layout], select [Modify] → select [Set color], change the Slip Ring to yellow, and change the Column to green in the same way, as shown in Fig. 15-8.

（5）在【布局】里在滑环上点击鼠标右键，选择【修改】→选择【设定颜色】，把滑环改为黄色，同理把立柱改为绿色，如图 15-8 所示。

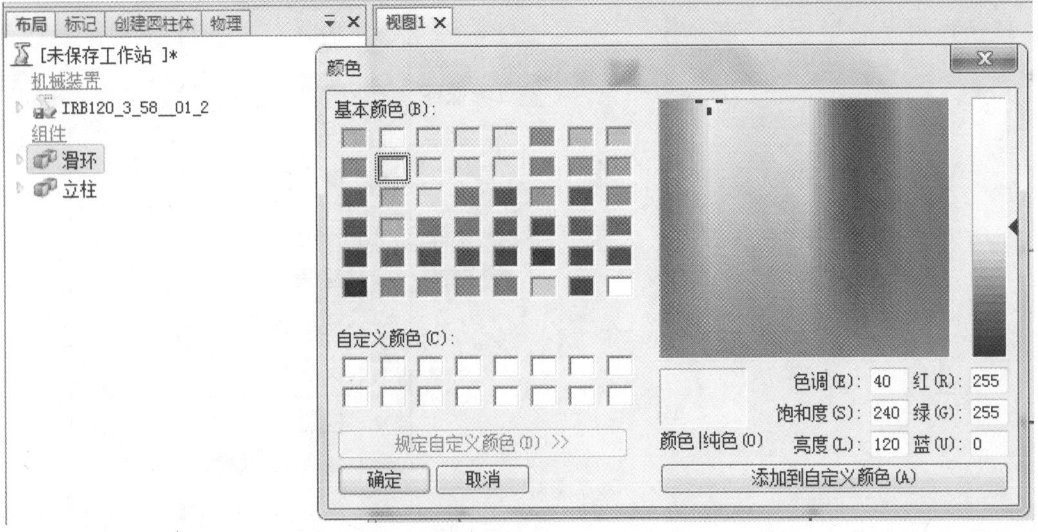

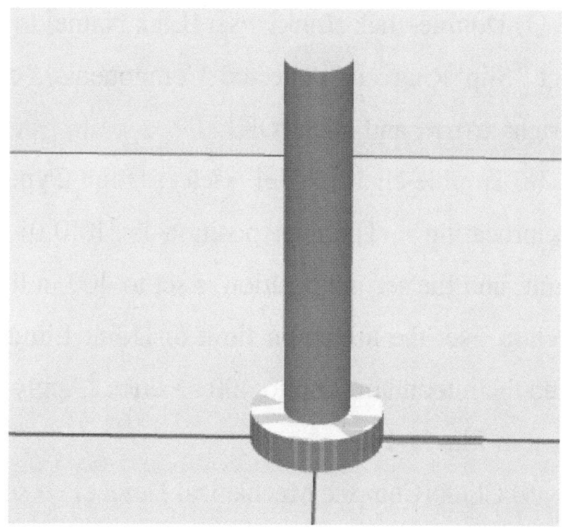

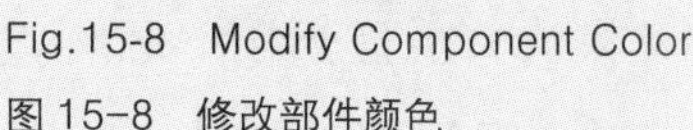

Fig.15-8 Modify Component Color

图 15-8 修改部件颜色

（6）点击【创建机械装置】→修改【机械装置模型名称】为“clyn”→【机械装置类型】设置为“设备”→双击【链接】→链接名称不改，【所选部件】选择“立柱”，勾选【设置为 BaseLink】，点击向右的箭头，点击【确定】，如图 15-9 所示。

(6) Click [Create Mechanical Device] → modify [Mechanical Device Model Name] to “clyn” → set [Mechanical Device Type] to “equipment” → double-click [Link] → do not change the link name, select “Column” in [Selected Component], check [Set as BaseLink], click the right arrow, and click [OK], as shown in Fig. 15-9.

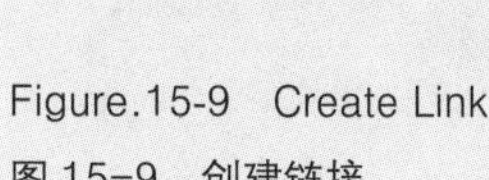

Figure.15-9 Create Link

图 15-9 创建链接

(7) Double-click [Link]→set [Link Name] to L2, select "Slip Ring" in [Selected Components], click the right arrow, and click [OK].

(8) Double-click [Node]→select [Joint Type] to "Reciprocating"→The first position is [0,0,0] by default, and the second position is set to 400 in the Z direction→set the minimum limit of [Joint Limit] to 0, and the maximum limit to 400 → click [Apply], as shown in Fig.15-10.

(9) Click [Compile Mechanical Device] → select "Manual Joint", "Select Component" → click the Slip Ring and hold on, swipe the mouse up and down to move the Slip Ring, as shown in Fig.15-11. So far, the simple mechanical device is created.

（7）鼠标左键双击【链接】→【链接名称】设置为 L2，[所选部件] 选择 "滑环"，点击向右的箭头，点击【确定】。

（8）鼠标左键双击【节点】→【关节类型】选择 "往复的"→第一个位置默认为【0，0，0】，第二个位置在 Z 方向上设置为 400→【关节限值】最小限值为 0，最大限值 400→点击【应用】，如图 15-10 所示。

（9）点击【编译机械装置】→选择 "手动关节" "选择部件" →鼠标左键点中滑环不放，上下移动鼠标就可以移动滑环，如图 15-11 所示。至此简单的机械装置就创建完成。

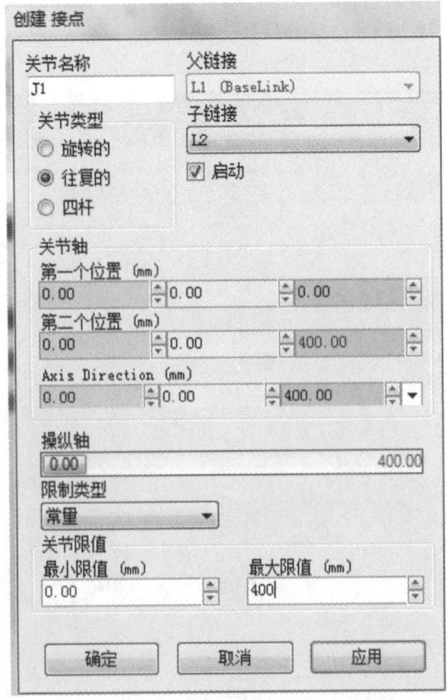

Fig.15-10 Create Contact

图 15-10 创建接点

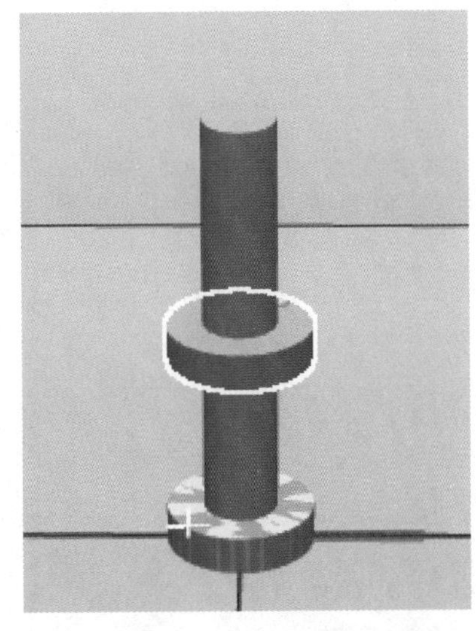

Fig. 15-11 Moving Slip Ring

图 15-11 移动滑环

微信扫一扫

教学视频

Lecture 15-2

A Modeling for Industrial Handling Robots and and Simulation Workstation(2)

15-2 工业搬运机器人的建模与仿真工作站（2）

我们在 15-1 机械装置的基础上，继续对机械装置进行设计，设计步骤如下：

（1）双击【创建机械装置】标签，点击【添加】，在【姿态名称】使用默认的名称“姿态1”，添加滑环的位置数据，将关节值设置为 0 的位置，单击“确定”，如图 15-12 所示。

Based on the mechanical device we learned in 15-1, we're going to continue to design the device. The design steps are as follows:

(1) Double-click the [Create Mechanical Device], click [Add], use the default name “Pose 1” in [Pose Name], add the position data of the slip ring, set the joint value to the position of 0, and click OK, as shown in Fig. 15-12.

Fig. 15-12 Modifypost

图 15-12 修改姿态

（2）重复上面的过程，点击“添加”，在“姿态名称”中输入“姿态 400”，添加滑环的定位位置的数据，将关节值设置为 400，单击“确定”。

(2) Repeat the above process, click “Add”, enter “Pose 400” in the “Pose Name”, add the location data of the slip ring, set the joint value to 400, and click “OK”.

(3) In the "Create Mechanical Device" dialog box, click "Set Transition Time" to set the movement time of the slip ring between the two positions, assuming that the transition time between "Pose 1" and "Pose 400" is set to 2 seconds, click "OK" when done, ,as shown in Fig. 15-13.

（3）在"创建机械装置"对话框中单击"设置转换时间"，设定滑环在两个位置之间的运动时间，假设"姿态 1"和"姿态 400"之间的转换时间设定为 2 秒，完成后单击"确定"，如图 15-13 所示。

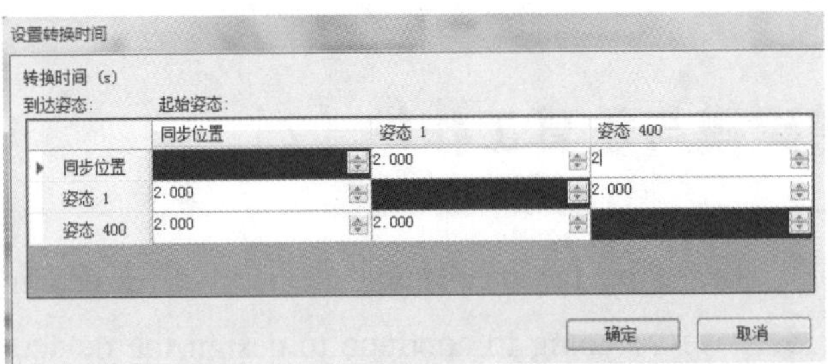

Fig. 15-13 Set transition time
图 15-13 设置转换时间

Now we are going to create the I/O signals and new events that trigger the action. The production steps are as follows:

(1) Select "controller" in the menu bar, select "configuration", click "I/O system", click "signal", right-click the signal on the right, and select "new signal... "to create a virtual signal, as shown in Fig. 15-14.

现在我们要创建触发动作的 I/O 信号和新的事件。制作步骤如下：

（1）在菜单栏中选择"控制器"，选择"配置"，单击"I/O system"，鼠标左键点击"signal"，在右边的信号上面点击鼠标右键，选择"新建 signal..."创建虚拟信号，如图 15-14 所示。

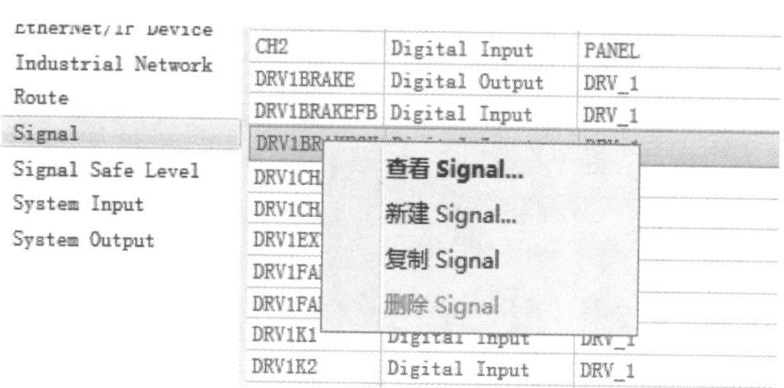

Fig. 15-14 Set I/O signal 1

图 15-14 设置 I/O 信号 1

（2）虚拟信号设置为“domove”，“Type of signal”设置为“Digital Output”，点击确定，系统要求重启，单击确定，并点击【重启】中的【重启动】，重启系统，如图 15-15 所示。

(2) Set the virtual signal to “domove” and “Type of signal” to “Digital Output”, click OK, the system requires restart, click OK, and click [Reboot] in [Restart] to restart the system, as shown in Fig. 15-15.

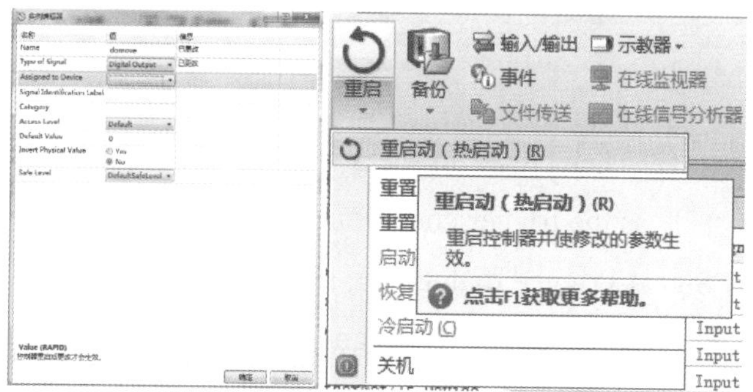

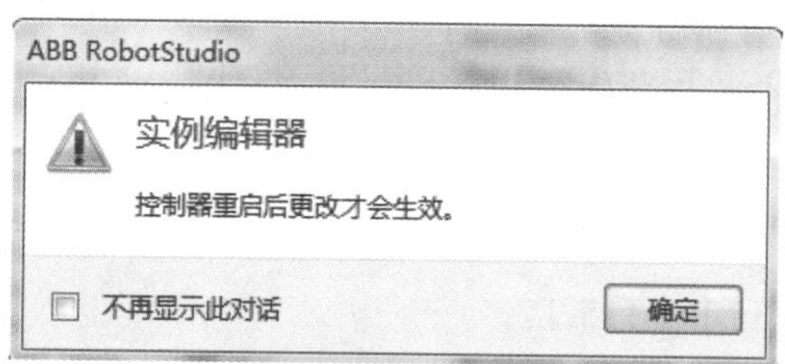

Fig. 15-15 Set I/O signal 2

图 15-15 设置 I/O 信号 2

下来我们要创建新的事件。制作步骤如下：

（1）点击菜单栏中的“仿真”→单击配置功能组右下角箭头→打开事件管理器窗口→单击任务窗格中的“添加”，打开创建新事件向导→选择“I/O 信号已更改”→单击“下一步”，如图 15-16 所示。

Next, we are going to create a new event. The production steps are as follows:

(1) Click “Simulation” in the menu bar→click the arrow in the lower right corner of the configuration function group→open the Event Manager window→ click “Add…” in the task pane to open the Create New Event Wizard→select “I/ O signal has been changed” → click “Next”,as shown in Fig. 15-16.

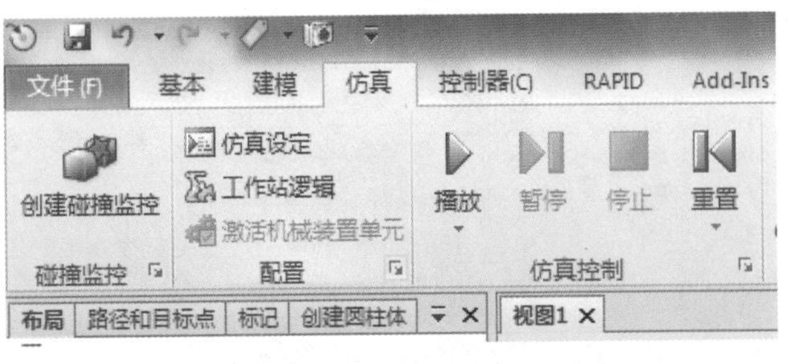

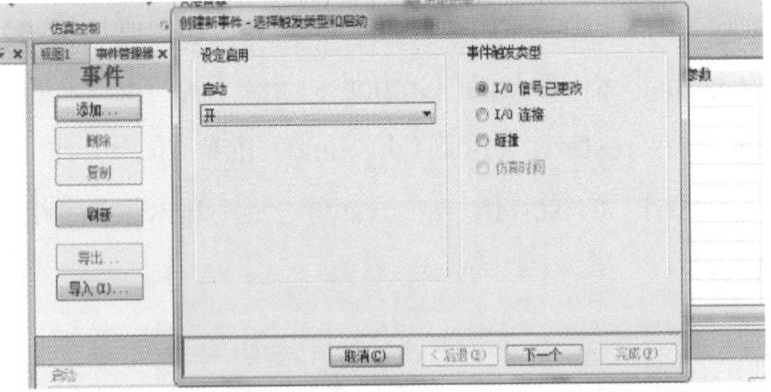

Fig. 15-16 Setsimulation event 1

图 15-16 设置仿真事件 1

(2) Then create an event with the "Signal is True ('1')" as the trigger condition. In the "New Event-I/O Signal Trigger" dialog box, select the newly created virtual signal "domove", "Signal is True ('1')" as the trigger condition, and click "Next". Set the action type to "Move the mechanical device to the pose", click "Next", select clyn for the mechanical device, select "Pose 400" for pose, and click Finish, as shown in Fig. 15-17.

（2）接着新建触发条件为"信号是 True（'1'）"事件。在"新建新事件-I/O 信号触发器"对话框中，选择新建的虚拟信号"domove"，触发条件为"信号是 True（'1'）"，单击"下一个"。设定动作类型为"将机械装置移至姿态"，单击"下一个"，机械装置选择 clyn，姿态为"姿态 400"，单击"完成"，如图 15-17 所示。

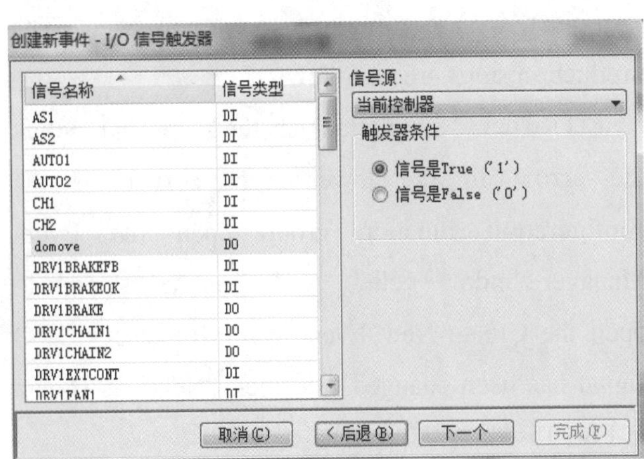

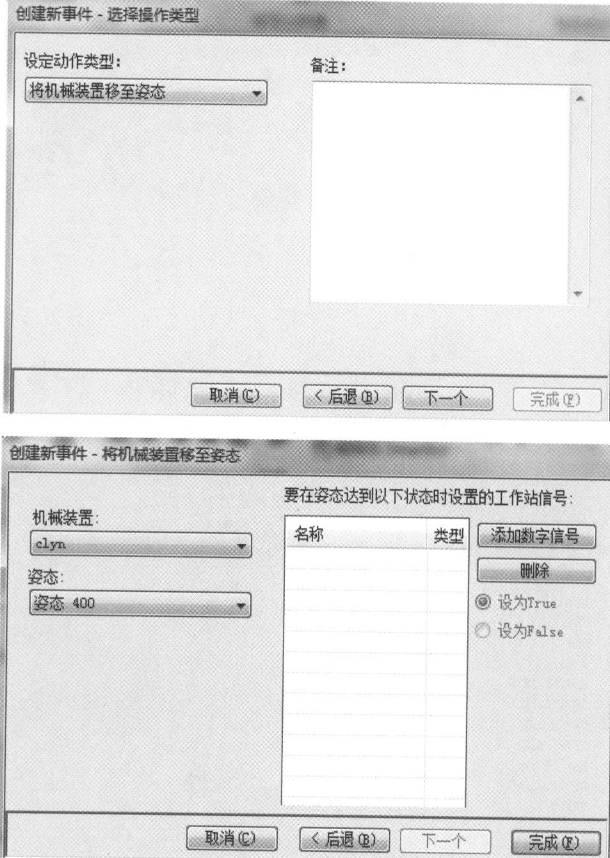

Fig. 15-17 Setsimulation event 2

图 15-17 设置仿真事件 2

(3) 接着新建触发条件为"信号是 False（'0'）"事件。过程与如上的类似，在"新建新事件-I/O 信号触发器"对话框中，选择新建的虚拟信号"domove"，触发条件为"信号是 False（'0'）"，单击"下一个"。设定动作类型为"将机械装置移至姿态"，单击"下一个"，机械装置选择 clyn，姿态为"姿态 1"，单击完成，如图 15-18 所示。

(3) Then create an event with the "signal is False ('0')" as the trigger condition. The process is similar to the above. In the "New Event-I/O Signal Trigger" dialog box, select the newly created virtual signal "domove", "Signal is False (' 0')" as the trigger condition, and click "Next". Set the action type to "Move the mechanical device to the pose", click "Next", select clyn for the mechanical device, select "Pose 1" for pose, and click "Finish", as shown in Fig. 15-18.

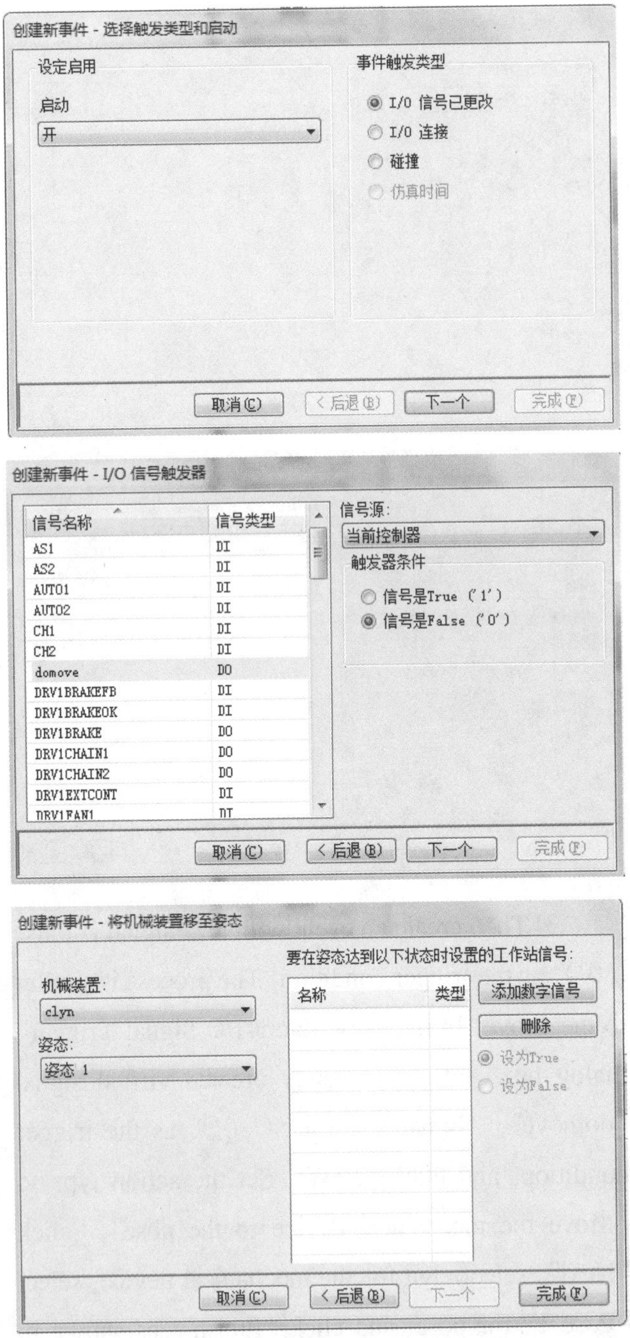

Fig. 15-18 Setsimulation event 3

图 15-18 设置仿真事件 3

(4) The events have been linked, and then we will make another program to let the robot control the device. In the menu bar "basic", select "path"→ select "empty path"→ "path_10" appears in "paths and steps" under "paths and target points"→ right-

（4）事件已关联好了，现在再编一个程序，让机器人去控制这个装置。在菜单栏"基本"里，选择"路径"→选择"空路径"→在"路径和目标点”下面的“路径和步骤”里面出现了“path_10”→在“path_10”上面点击鼠标右键，选择插入“逻辑指令”，如图15-19所示。

click the “path_10” and select to insert “Logical Instructions”, as shown in Fig. 15-19.

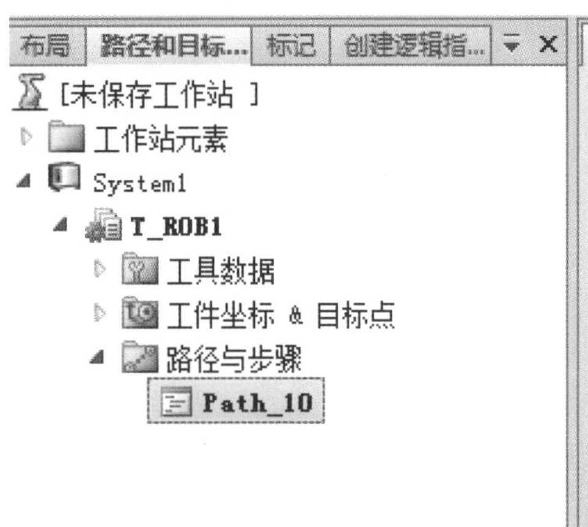

Fig. 15-19　Empty path path_10
图 15-19　空路径 path_10

（5）在“指令模板”里选择“Set”，信号是默认的 domove，点击“创建”，如图 15-20 所示。

(5) Select “Set” in the “Instruction Template”, the default “domove” as the signal, and click “Create”, as shown in Fig. 15-20.

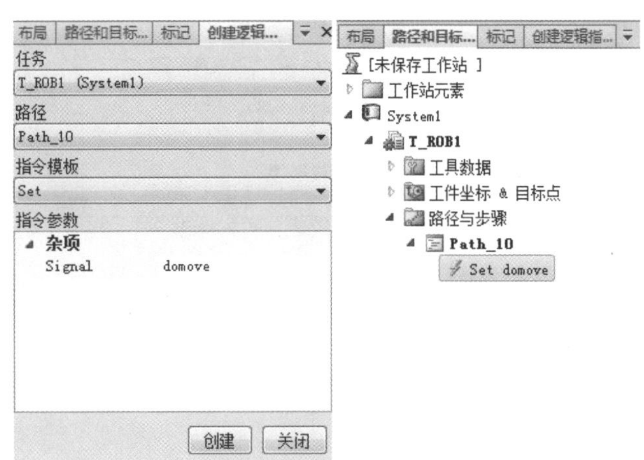

Fig. 15-20　Empty path path_10
图 15-20　空路径 path_10

（6）然后再等待 5 秒，在“指令模板”里选择“WaitTime”，Time 设置为 5，点击创建，然后再复位 domove 信号，在“指令模板”里选择“Reset”，信号默认是 domove，再复制“WaitTime 5”，并在

(6) Then wait another 5 seconds, select “WaitTime” in the “Instruction Template”, set the Time to 5, click Create, and then reset the domove signal, select “Reset” in the “Instruction Template”, the default signal is domove, then copy “WaitTime 5”, right-click the “Reset domove” command, and select Paste, as shown in Fig. 15-21.

"Reset domove" 指令上点击鼠标右键，选择"粘贴"，如图 15-21 所示。

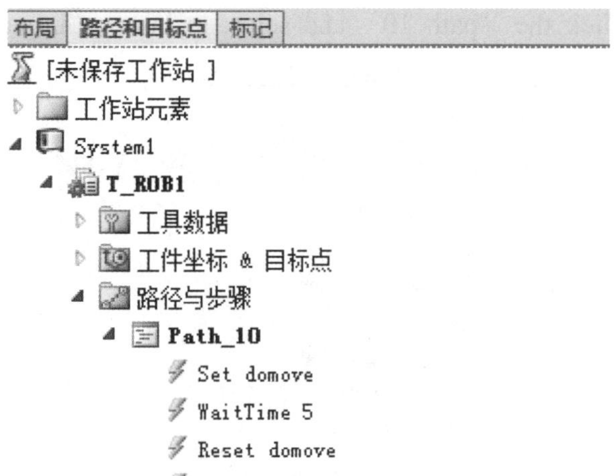

Fig. 15-21 Effect of program

图 15-21 程序效果

(7) Finally, synchronize the program to the robot, select "Sync", and click OK, as shown in Fig. 15-22.

（7）最后把程序同步到机器人中，选择"同步"，点击"确定"，如图 15-22 所示。

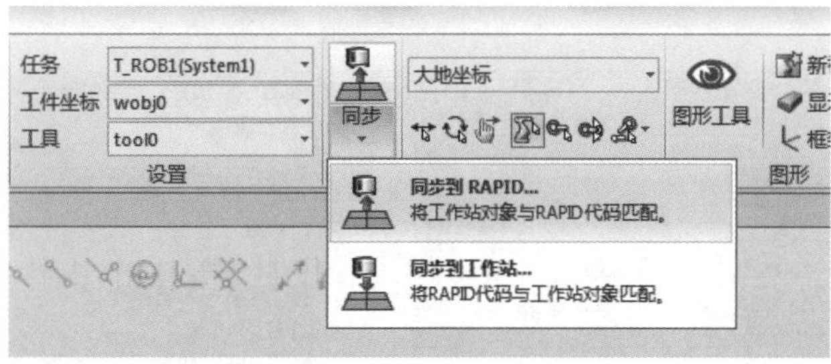

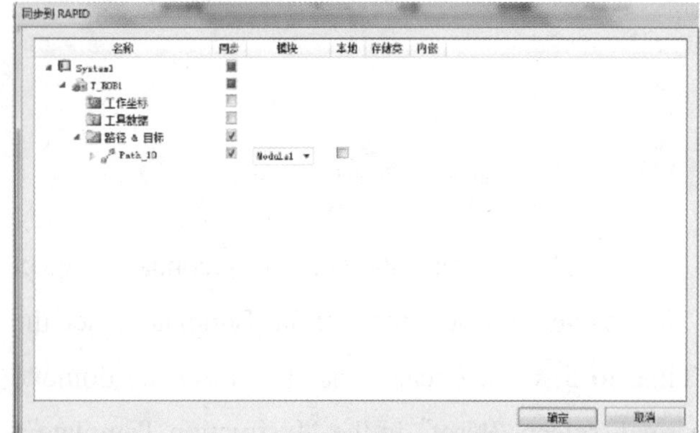

Fig. 15-22 Synchronize the program to the robot

图 15-22 程序同步到机器人

（8）最后进入点选择 path_10，回到视图 1，点击“播放”，可以看到程序运行效果，如图 15-23 所示。

(8) Finally, enter the point and select path_10, return to view 1, click “Play”, and you can see the program running effect, as shown in Fig. 15-23.

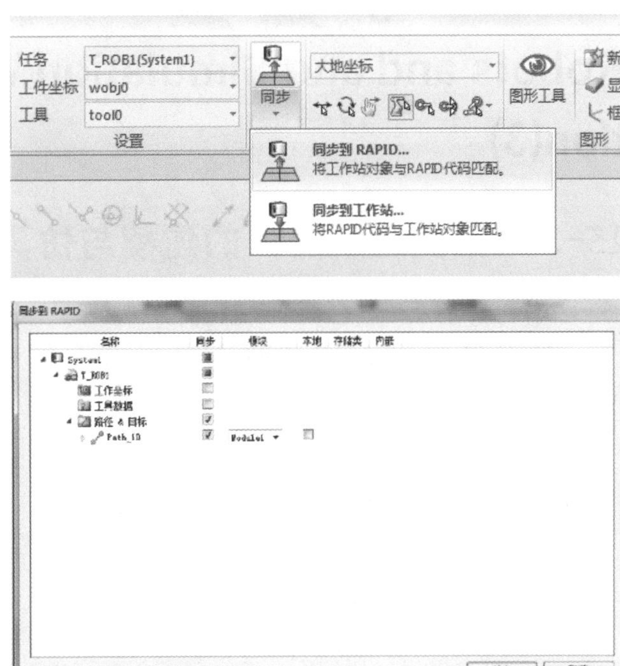

Fig. 15-23 program running effect

图 15-23 仿真运行效果

这一节我们在 RobotStudio 软件中利用事件管理器和触发器，制作滑环和立柱机构的动画效果。

In this part we use the event manager and triggers in RobotStudio software to make animation effects of slip ring and column mechanism.

微信扫一扫

教学视频

Lecture 15-3

A modeling for Industrial Handling Robots and and Simulation Workstation(3)

15-3 工业搬运机器人的建模与仿真工作站（3）

This part we will use the event manager to achieve the transmission of objects on the conveyor chain.

First, we need to create a sliding table and a slider device. The production method is as follows:

(1) As mentioned before, create a robot control system and design a rectangular body with a length of 2000, a width of 500, and a height of 100, the color is set to yellow, and the name is changed to the sliding table;

(2) In the same way, design a rectangular body, the corner point parameter Y is 150, Z is 100, the length is 200, the width is 200, the height is 100, the color is set to green, and the name is changed to slider. The effect is shown in Fig. 15-24;

这一部分我们要利用事件管理器实现传送链上物体的传送。

首先，我们要创建一个滑台和滑块装置，制作方法如下所示：

（1）如之前所述，创建一个机器人控制系统，并设计一个矩形体，长度为 2000，宽度为 500，高度为 100，颜色设置为黄色，名称改为滑台；

（2）同理，设计一个矩形体，角点参数 Y 为 150，Z 为 100，长度为 200，宽度为 200，高度为 100，颜色设置为绿色，名称改为滑块。效果如图 15-24 所示；

Fig.15-24 Sliding table and slider device

图 15-24 滑台与滑块装置

（3）点击菜单栏“建模”里的“移动”，点击滑块，将滑块沿 X 轴正方向移至 1200mm 左右处，如图 15-25 所示。

(3) Click “Move” in the menu bar “Modeling”, click the slider, click the slider, and move the slider to about 1200mm along the positive direction of the X axis, as shown in Fig. 15-25.

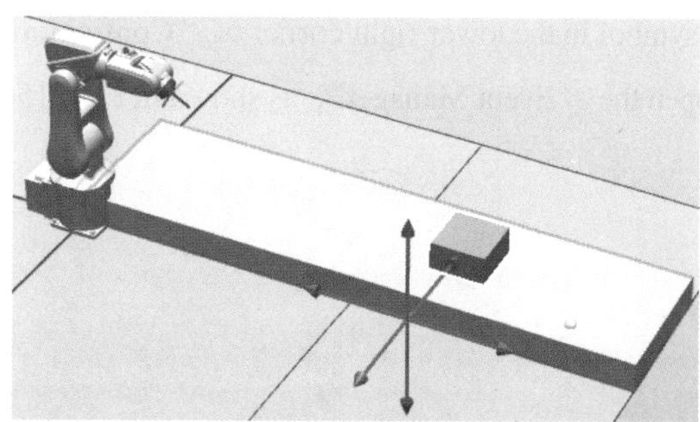

Fig.15-25 Move the slider along the positive direction of the X axis

图 15-25 沿 X 轴正方向移动滑块

接下来，我们来设计触发信号和新的事件。这里滑块在滑台上沿着 X 轴负方向运动，实际上是利用视觉误差让物体从初始位置移动三次到终点位置，滑块位置的每次改变都是利用一个输出信号置 1 来实现，然后再进行适当的延时，所以这里需要有三个数字量的输出信号，信号的设置和事件的设置如下所示：

Next, let's design trigger signals and new events. Here the slider moves in the negative direction of the X-axis on the sliding table. Actually, the object is moved from the initial position to the end position three times by visual error. Each change of the position of the slider is realized by setting an output signal to 1. Then perform an appropriate delay, so three digital output signals are needed here. The signal settings and event settings are as follows:

（1）点击工具栏中的“控制器”，选择“配置管理器”，选择“I/O signal”，选择“signal”，单击鼠标右键，单击“新建 signal...”，创建三个虚拟信号，名称分别为 domove1，domove2 和 domove3，如图 15-26 所示。

(1) Click “Controller” in the toolbar, select “Configuration Manager”, select “I/O signal”, select “signal”, right-click the mouse, click “New signal...” to create three virtual signals named domove1, domove2 and domove3, as shown in Fig. 15-26.

类型
Access Level
Cross Connection
Device Trust Level
EtherNet/IP Command
EtherNet/IP Device
Industrial Network
Route
Signal
Signal Safe Level

Name	Type of Signal
AS1	Digital Input
AS2	Digital Input
AUTO1	Digital Input
AUTO2	Digital Input
CH1	Digital Input
CH2	Digital Input
domove1	Digital Output
domove2	Digital Output
domove3	Digital Output
DRV1BRAKE	Digital Output

Fig.15-26 Create three digital signals

图 15-26 创建三个数字量信号

(2) Click "Restart" in "Controller" in the menu bar and select "Hot Start" to restart the control system to make the new signal effective.

（2）点击菜单栏的“控制器”里的“重启”，选择“热启动”，重启控制系统，使新建的信号生效。

(3) Click "Simulation" in the menu bar, click the symbol in the lower right corner of "Configuration" to open the "Event Manager", as shown in Fig. 15-27.

（3）点击菜单栏的“仿真”，点击“配置”的右下角的符号，打开“事件管理器”，如图 15-27 所示。

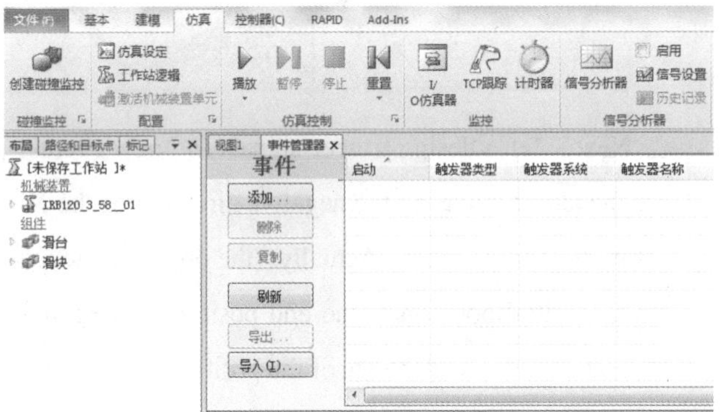

Fig.15-27 Open the event manager

图 15-27 打开事件管理器

(4) Click "Add", select "I/O signal has been changed" in "Event trigger type", click "Next", select domove1 for the signal, select "Signal is True (' 1')" in "Trigger condition", click "Next", select "Move object" in "Set action type", click "Next", select "Slider" in "Object to move", the first blank of the position is the position in the X direction, which is filled with 800 here, and the second blank is filled with 0. In the same way, set the position of the slider when domove2 and domove3 are triggered, and fill 400 and 0 in the X direction, and 0 in the Y direction, as shown in Fig. 15-28.

（4）点击“添加”，“事件触发类型”选择“I/O 信号已更改”，点击“下一个”，信号选择 domove1，“触发器条件”选择“信号是 True（‘1’）”，点击“下一个”，“设定动作类型”选择“移动对象”，点击“下一个”，“要移动的对象”选择“滑块”，位置的第一个空是 X 方向上的位置，这里填 800，第二个空填 0。同理，设置 domove2 和 domove3 触发时滑块的位置，分别在 X 方向上的位置填 400 和 0，Y 方向上填 0，如图 15-28 所示。

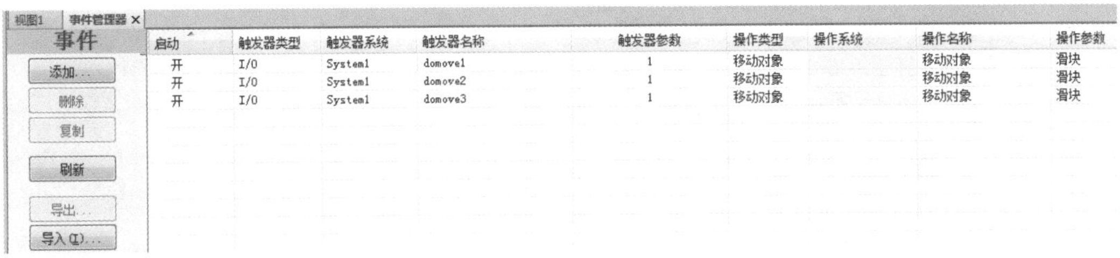

启动	触发器类型	触发器系统	触发器名称	触发器参数	操作类型	操作系统	操作名称	操作参数
开	I/0	System1	domove1	1	移动对象		移动对象	滑块
开	I/0	System1	domove2	1	移动对象		移动对象	滑块
开	I/0	System1	domove3	1	移动对象		移动对象	滑块

Fig 15-28　Setting up the event manager

图 15-28　设置事件管理器

（5）点击菜单栏中的“基本”，点击“路径”，选择“空路径”，出现了 path_10。

（6）在 path_10 上点击鼠标右键，选择“插入逻辑指令”，在“指令模块”里选择“Set Default”，指令参数选择 domove1，点击“创建”。

（7）在“指令模板”里选择“WaitTime Default”，Time 设置为 2，延时 2 秒。

（8）同理复位 domove1，再置位 domove2，再延时 2 秒，再复位 domove2。

（9）置位 domove3，延时 2 秒，复位 domove3。

（10）选择“同步”，点击“同步到 RAPID...”，点击“确定”，然后选择“仿真设定”，进入点选择“path_10”，回到“视图 1”，点击“播放”，可看到程序的运行效果，如图 15-29 所示。

(5) Click “Basic” in the menu bar, click “Path”, select “Empty path”, and path_10 appears.

(6) Right-click the path_10, select “Insert Logical Instruction”, select “Set Default” in “Instruction Module”, select domove1 as the instruction parameter, and click “Create”.

(7) Select “WaitTime Default” in “Instruction Template”, set Time to 2 and delay 2 seconds.

(8) In the same way, reset domove1, set domove2 again, delay another 2 seconds, and then reset domove2.

(9) Set domove3, delay 2 seconds, and reset domove3.

(10) Select “Sync”, click “Sync to RAPID...”, click “OK”, then select “Emulation Settings”, enter the point and select “path_10”, return to “View 1”, click “Play”, and the effect of the program can be seen, as shown in Fig. 15-29.

Fig.15-29 Insert logical instruction
图 15-29 插入逻辑指令

We created a mechanical device of a slider and a sliding table in RobotStudio by using the event manager, and controlled the movement of the slider through three signals. If you want the slider to move more continuously, you can use the same method to set the motion position and digital output signal.

我们利用事件管理器在RobotStudio软件中创建了一个滑块与滑台的机械装置，并且通过三个信号来控制滑块的动作。如果想让滑块运动得更加连续，可以用同理的方法多设置运动位置和数字输出信号。

微信扫一扫

教学视频

Lecture 15-4
A Modeling for Industrial Handling Robots and and Simulation Workstation(4)

15-4 工业搬运机器人的建模与仿真工作站（4）

这一部分需要让滑块运动到终点位置时机器人去抓取滑块，并将滑块放下。

制作方法如下所示:

（1）给机器人设计一个抓取工具。点击菜单栏中“建模”，点击“固体”，点击“圆柱体”，半径输入 50，高度输入 200，点击“创建”。

（2）将圆柱体的名字改为“工具”，并将颜色改为红色。

（3）鼠标左键点击“工具”不放，移动光标到机器人，松开鼠标左键，将工具装到机器人的法兰盘上，如图 15-30 所示。

（4）设定一个信号，当信号为 1 时，工具把物体抓起来；当信号为 0 时，工具把物体放下。点击菜单栏中的“控制器”，

This part we will make the robot grab the slider and put it down when the slider moves to the end position.

The production method is as follows:

(1)Design a grabbing tool for the robot. Click “Modeling” in the menu bar, click “Solid”, click “Cylinder”, enter 50 for radius, 200 for height, and click “Create”.

(2)Change the name of the cylinder to “tool” and change the color to red.

(3)Click the “Tool” and hold on, move the cursor to the robot, release the left mouse button, and install

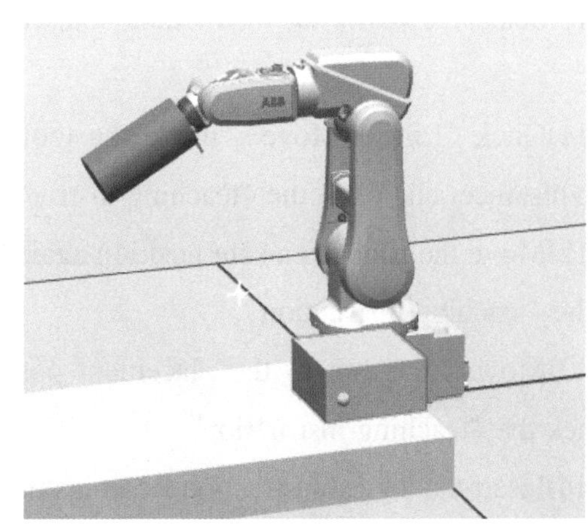

Fig.15-30 Tools installed on the robot

图 15-30 工具装到机器人上

the tool on the flange of the robot,as shown in Fig. 15-30.

(4)Set a signal. When the signal is 1, the tool grabs the object; when the signal is 0, the tool puts the object down. Click "controller" in the menu bar, select "configuration", select "I/O system", select "signal", create a new digital output signal "doTool", and then restart the controller.

(5)Click the last line of program in path_10 in "path and target point" to continue programming. Change the robot motion command to MoveJ and the speed to fine.

(6)Click "manual joint", click the first joint of the robot, turn the robot to a suitable position, and then turn the fifth axis down as the first point, click "teaching instruction".

(7)Click "Linear Move", move the tool to above of the object, and click "teaching instruction".

(8) Put the tool in contact with the object and click the "teaching instruction".

(9)Right-Click the path_10, select "insert logical instruction...", select the set instruction, the instruction parameter is doTool, and click "Create".

(10)Select the WaitTime instruction, and set Time to 1.

(11)Click "Linear Move", move the tool up a certain distance, and click the "teaching instruction".

(12)Move the robot to a safe position again, and click the "teaching instruction".

(13)Move the robot to the placement position, and click the "teaching instruction".

(14)Insert the logical instruction Reset instruction,

选择"配置"选择"I/O system",选择"signal",新建数字输出信号"doTool",然后重启控制器。

（5）点击"路径和目标点"里的 path_10 里的最后一行程序，继续编程。把机器人运动指令换成 MoveJ，速度换成 fine。

（6）点击"手动关节"，点击机器人的第一个关节，把机器人转到合适的位置，再把第五轴转下来，作为第一个点，点击"示教指令"。

（7）点击"线性移动"，把工具移动到物体正上方，点击"示教指令"。

（8）把工具和物体接触，点击"示教指令"。

（9）在 path_10 上点击鼠标右键,选择"插入逻辑指令",选择 set 指令，指令参数为 doTool，点击"创建"。

（10）选择 WaitTime 指令，Time 设置为 1。

（11）点击"线性移动"，把工具往上移动一段距离，点击"示教指令"。

（12）把机器人再移至安全位置，点击"示教指令"。

（13）把机器人移至放置位置，点击"示教指令"。

（14）插入逻辑指令 Reset 指令，把 doTool 置 0，将物体放下，再输入 WaitTime 指令，Time 设为 1。

（15）把机器人往上移一段距离，点击“示教指令”。

（16）点击“同步”，选择“同步到 RAPID…”并把物体移到X方向上大于800的位置。

（17）点击“仿真”，打开“事件管理器”，点击“添加”，“事件触发类型”选择“I/O 信号已更改”，点击“下一个”，信号选择 doTool，“触发器条件”选择“信号是 True（‘1’）”，点击“下一个”，设定“动作类型”选择“附加对象”，点击“下一个”，“附加对象”选择“滑块”，“安装到”选择“工具”，勾选“保持位置”，点击“完成”。

（18）再点击“添加”，“事件触发类型”选择“I/O 信号已更改”，点击“下一个”，信号选择 doTool，“触发器条件”选择“信号是 False（‘0’）”，点击“下一个”，“设定动作类型”选择“提取对象”，点击“下一个”，“提取对象”选择“滑块”，“提取于”选择“工具”，点击“完成”。

（19）点击“仿真设定”，查看进入点是否“path_10”，点击“播放”，程序开始运行，机器人开始运动。

set doTool to 0, put the object down, then input the WaitTime instruction, and set Time to 1.

(15) Move the robot up a certain distance, and click the “teaching instruction”.

(16) Click “Sync”, select “Sync to RAPID…” and move the object to a position greater than 800 in the X direction.

(17) Click “Simulation”, open “Event Manager”, click “Add”, select “I/O signal has been changed” for “Event trigger type”, click “Next”, select doTool for signal, select “Signal is True(’1’)” in “Trigger condition”, click “Next”, set “Action Type”, select “Additional Object”, click “Next”, select “Slider” in “Additional Object”, select “Tools” in “Install to”, check “Keep Position”, and click “Finish”.

(18) Then click “Add”, select “I/O signal has been changed” in “Event Trigger Type”, click “Next”, select doTool for signal, select “Signal is False(’0’)” in “Trigger Condition”, Click “Next”, select “Extract Object” in “Set Action Type”, click “Next”, select “Slider” in “Extract Object”, select “Tool” in “Extract from”, and click “Finish”.

(19)Click “Simulation Settings” to check whether the entry point is “path_10”, click “Play”, the program starts to run, and the robot starts to move.

微信扫一扫

教学视频

Lecture 15-5

A Modeling for Industrial Handling Robots and and Simulation Workstation(5)

15-5 工业搬运机器人的建模与仿真工作站（5）

The methodof creating a conveyor chain transmission effect in RobotStudio to match the robot's actions.

The current palletizing workstation requires several operating effects. The first is the operation effect of the conveyor chain, the second is the grabbing and putting-down effect of the fixture, and the third is the connection between the conveyor chain, the fixture and the robot. First of all, we need to implement the conveyor chain operation effect. This effect includes several features. The front end of the conveyor chain can automatically generate materials, and the materials can automatically run along the conveyor chain. When the materials move to the end, they stop and send out an in-position signal. When the robot picks up materials, the front end of the conveyor chain can automatically generate materials for the next cycle. The realization of these animation effects requires the help of SMART components. The steps are as follows:

(1) Open RobotStudio6 software and create a new

在 RobotStudio 中创建一个输送链传动效果以配合机器人动作的方法。

当前码垛工作站需要几个运行效果，首先是输送链运行效果，第二是夹具的抓取、放下效果，第三是输送链、夹具和机器人之间的联系。首先我们要做输送链运行效果，这个效果包括几个特征，输送链前端能自动生成物料，物料能自动沿着输送链运行，物料运动到末端后停止并发出到位信号，当机器人拾取物料以后，输送链前端能自动生成物料进行下一个循环。这些动画效果的实现需要借助SMART组件。制作步骤如下：

（1）打开 RobotStudio6 软件，新建一个机器人系统。

（2）在菜单栏“HOME”中，点击“导入模型库”，点击

"输送链"和"Euro Pallet",并把它们分别放在机器人两侧的合适的位置。

(3)在菜单栏"建模"中,点击"导入几何体",点击"浏览几何体",选中吸盘所在的路径,点击"打开",并用鼠标左键点击工具不放,拖至机器人上松开鼠标左键,将工具安装到机器人末端法兰上,如图15-31所示。

robot system.

(2) In the menu bar "HOME", click "Import Model Library", click "Conveyor Chain" and "Euro Pallet", and place them in the appropriate positions on both sides of the robot.

(3) In the menu bar "Modeling", click "Import Geometry", click "Browse Geometry", select the path where the suction cup is located,and click "Open". Click the tool and hold on, drag it to the robot, and release the left mouse to install the tool on the end flange of the robot, as shown in Fig. 15-31.

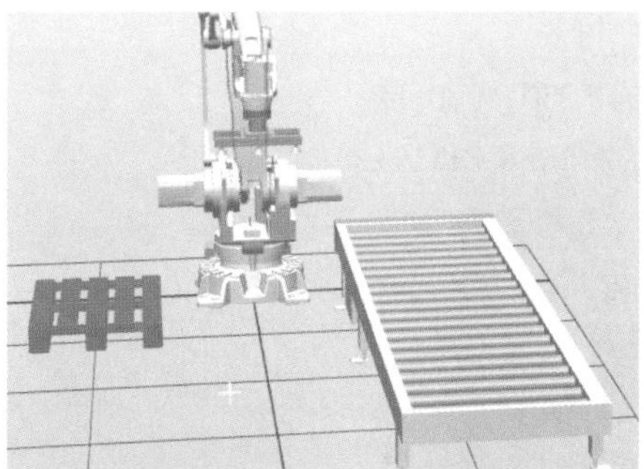

Fig.15-31 Layout of the Palletizing Workstation
图15-31 码垛工作站的布局图

(4)在输送链前端画一个长方体箱子,长度为800mm,宽度为500mm,高为400mm,颜色设置为绿色,名字重命名为箱子,并把箱子放在输送链的中间,如图15-32所示。

(4) Draw a cuboid box at the front end of the conveyor chain with a length of 800mm, a width of 500mm, and a height of 400mm. The color is set to green, the name is renamed to the "box", and the box is placed in the middle of the conveyor chain, as shown in Fig. 15-32.

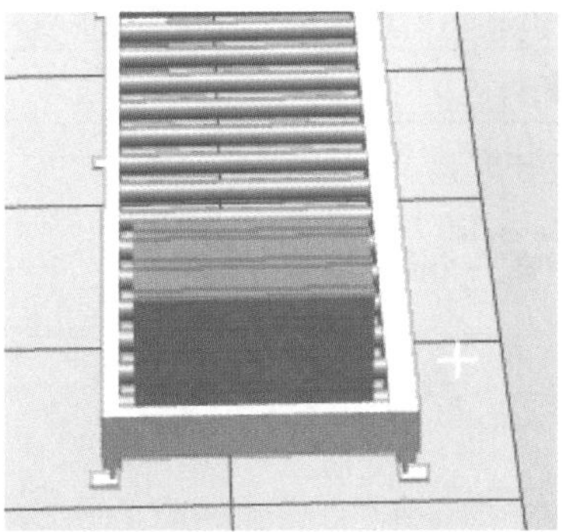

Fig.15-32 Box Placement Diagram
图15-32 箱子放置图

(5) Right-click on the box in "Layout", click "Modify", and click "Set Local Origin".

(6) Select "Modeling" in the menu bar, select "SMART component", click the newly created SMART component, rename it to "SC_Infeeder", and then add various subcomponents inside, as shown in Fig. 15-33.

（5）在"布局"中在箱子上点击鼠标右键，点击"修改"，点击"设定本地原点"。

（6）在菜单栏选择"建模"，选择"SMART 组件"，点击新建的 SMART 组件，给它重命名为"SC_Infeeder"，如图 15-33 所示。

Fig.15-33 Create SMART Component

图 15-33 创建 SMART 组件

(7) Click "Add Component" in the SMART component "SC_Infeeder", click "Action", and click "source", as shown in Fig. 15-34.

（7）点击 SMART 组件"SC_Infeeder"里的"添加组件"，点击"动作"，点击"source"，如图 15-34 所示。

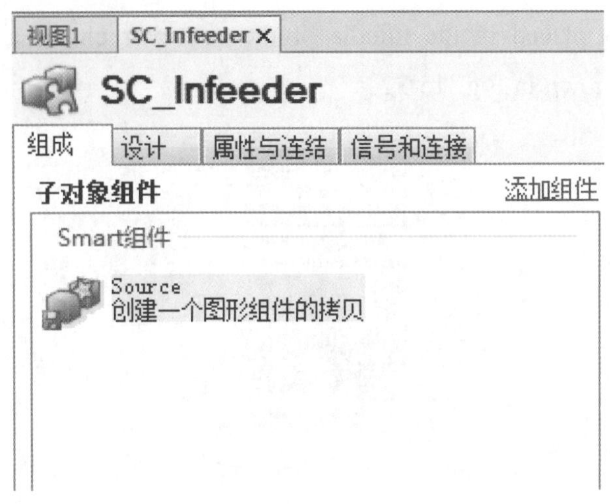

Fig.15-34 Add source Component

图 15-34 添加 source 组件

（8）Source 组件里的产品源是要放置的物料，即箱子，copy 和 parent 也置空，position 是每次生成的拷贝在空间中的空间信息。因为前面已经将箱子的位置设定为原点位置，所以这里全部设置为 0，trasient 先不勾选点击确定，如图 15-35 所示。

(8) The product source in the Source component is the material to be placed, that is, the box, the copy and parent are also left blank, and the position is the spatial information of each generated copy in the space. The position of the box has been set as the origin position, so here all settings are set to 0. Trasient is not checked first, and click "OK", as shown in Fig. 15-35.

Fig.15-35　Properties of the source Component

图 15-35　source 组件的属性

（9）点击"添加组件"，选择"本地"，点击"LinearMover"。

(9)Click "Add Component", select "Local", and click "LinearMover".

（10）点击"添加组件"，点击"其他"，点击"Queue"。

(10) Click "Add Component", click "Other", and click "Queue".

（11）对"LinearMover"组件进行设置，Object 选择"SC_Infeeder/Queue"，Direc-

(11)Set the "LinearMover" component, select "SC_Infeeder/Queue" in Object, enter -1 for the first space of Direction, Speed is set to 300, and the signal Execute is set to start, as shown in Fig. 15-36.

tion 的第一个空输入-1, Speed 设为 300,信号 Execute 设为启动,如图 15-36 所示。

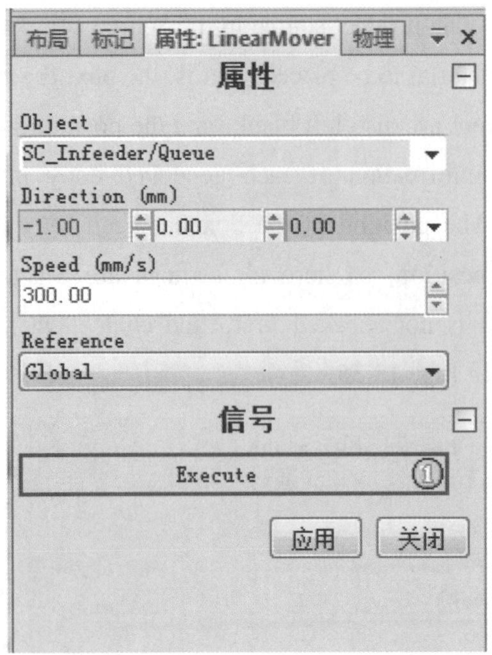

Fig.15-36 LinearMover Components

图 15-36 LinearMover 组件

(12)Click "Add Component", click "Sensor", and click "PlaneSensor" to create a surface sensor.

(13)There is no block on the original conveyor chain, so it is difficult to install the sensor. Therefore, it is necessary to set a block, click the "Modeling" in the menu bar, and click the "Solid". A block is designed at the end of the conveyor chain, and its color is set to red , as shown in Fig. 15-37.

(12) 点击"添加组件",点击"传感器",点击"PlaneSensor"。

(13)点击菜单栏中的"建模",点击"固体",在输送链的末端设计一个挡块,并把它的颜色设置为红色,如图 15-37 所示。

Fig.15-37 Design of the Block

图 15-37 挡块的设计

（14）鼠标右键点击 “PlaneSensor”，点击 “属性”，点击 “捕捉末端” 工具，点击 Origin 的第一个空，再点击红色挡块的左上角，这时 Origin 的坐标就写入到指定的空格里了，Axis1 的第一个空为 0，第二个空为 900，第三个空为 0，Axis2 的一个空位 0，第二个空为 0，第三个空为−50，如图 15−38 所示。

(14)Right-click the “PlaneSensor”,click “Properties”, click the “Capture End” tool, click the first space of Origin, and then click the upper left corner of the red block, then the coordinates of Origin will be written into the specified space. Here, the first space of Axis1 is 0, the second space is 900, and the third space is 0. The first space of Axis2 is 0, the second space is 0, and the third space is −50, as shown in Fig. 15-38.

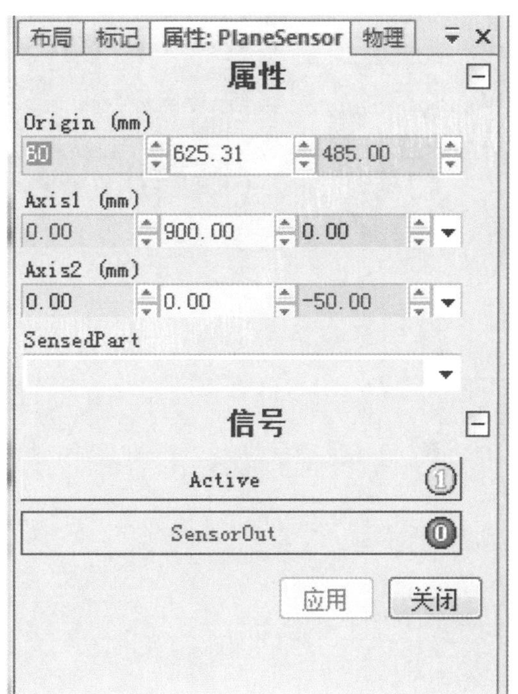

Fig.15-38 Design of the PlaneSensor

图 15−38 PlaneSensor 的设计

（15）点击两次 PlaneSensor 里的 Active，查看有无物体被检测，结果发现挡块被检测到，将被检测到的物体的 “可由传感器检测” 前面的钩去掉，如图 15−39 所示。

(15) Click Active in PlaneSensor twice to check whether there is an object detected, and it turns out that the block is detected. Remove the check mark in front of the detected object “can be detected by the sensor”, as shown in Fig. 15-39.

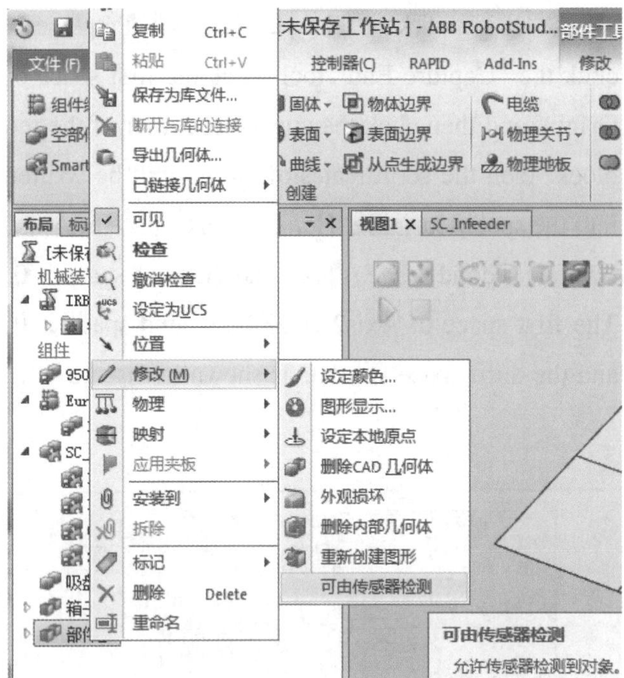

Fig.15-39 Remove the check mark before detection by the sensor

图 15-39 去掉可由传感器检测前的钩

Lecture 15-6

A Modeling for Industrial Handling Robots and and Simulation Workstation(6)

微信扫一扫

教学视频

15-6 工业搬运机器人的建模与仿真工作站（6）

输送链信号的联结方法。

在上节课制作的基础上，制作方法如下所示:

（1）在“属性与连结”下面点击“添加连结”，“源对象”选择“source”，“源属性”选择“copy”，“目标对象”选择“Queue”，“目标属性或信号”选择“Back”，如图 15-40 所示。

The method of connection of the conveyor chain signals.

Based on the production in the previous lesson, the steps are as follows:

(1)Click “Add Link” under “Attributes and Links”, select “source” for “Source Object”, select “copy” for “Source Attribute”, select “Queue” for “Target Object”, and select “Back” for “Target Attribute or Signal”, as shown in Fig. 15-40.

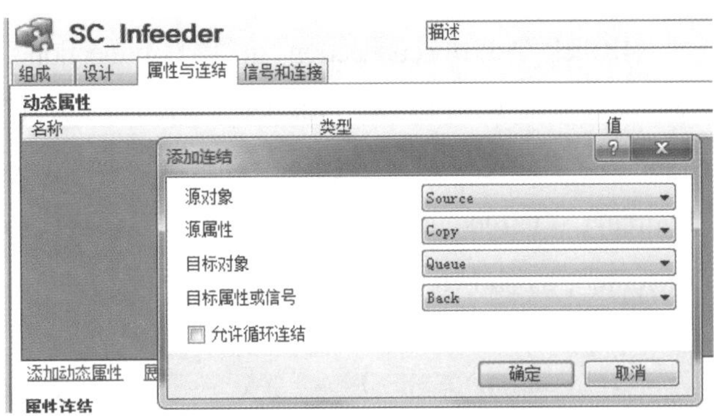

Fig.15-40 Setting of Properties and Bindings

图 15-40 属性与连结的设置

（2）点击“信号和连接”下面的“添加 I/O Signals”，添加一个启动信号，“信号类型”设为“数字输入”，“信号名称”设为“StartCNV”，如图 15-41 所示。

(2) Click “Add I/O Signals” under “Signals and Connections” to add a start signal. Set “Signal Type” to “Digital Input” and “Signal Name” to “StartCNV”, as shown in Fig. 15-41.

Fig. 15-41 Add Start Signal

图 15-41 添加启动信号

(3) Click “Add I/O Signals”, set “Signal Type” to “Digital Output”, “Signal Name” to “BoxInPos”, and click “OK”, as shown in Fig. 15-42.

（3）点击“添加 I/O Signals”，“信号类型”设置为“数字输出”，“信号名称”设置为“BoxInPos”，点击“确定”，如图 15-42 所示。

添加I/O Signals
信号类型
DigitalOutput
自动复位
信号数量
1
信号名称
BoxInPos
开始索引
0
步骤
1
信号值
0
最小值
0.00
最大值
0.00
描述
隐藏
只读
确定
取消

Fig.15-42 Add BoxInPos Signal

图 15-42 添加 BoxInPos 信号

(4)Click “AddI/O connection” in “I/O Connection”, “Source Object” is set to “SC_Infeeder”, “Source Signal” is set to “StartCNV”, “Target Object” is set to “Source”, “Target signal or attribute” is set to “Execute”, as shown in Fig. 15-43.

（4）点击“I/O 连接”里的“添加 I/O connection”，“源对象”设置为“SC_Infeeder”，“源信号”设置为“StartCNV”，“目标对象”设置为“Source”，“目标信号或属性”设置为“Execute”，如图 15-43 所示。

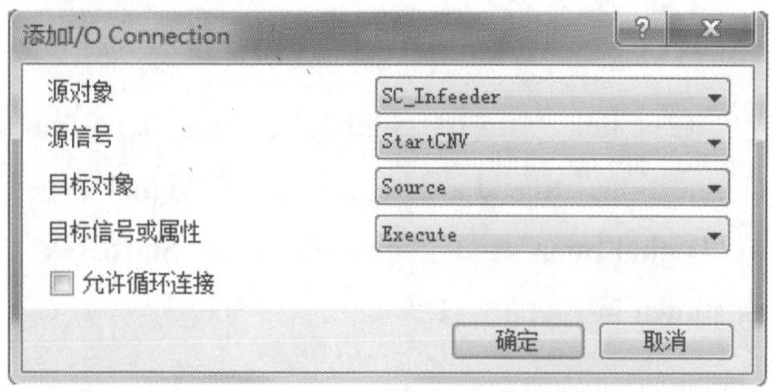

Fig.15-43 Add Signal Connection

图 15-43 添加信号连接

（5）点击“添加 I/O Connection”，“源对象”设置为“Source”，“源信号”设置为“Executed”，“目标对象”设置为“Queue”，“目标信号或属性”设置为“Enqueue”，如图15-44所示。

(5)Click “Add I/O Connection”, “Source Object” is set to “Source”, “Source Signal” is set to “Executed”, “Target Object” is set to “Queue”, and “Target Signal or Attribute” is set to “ Enqueue”, as shown in Fig. 15-44.

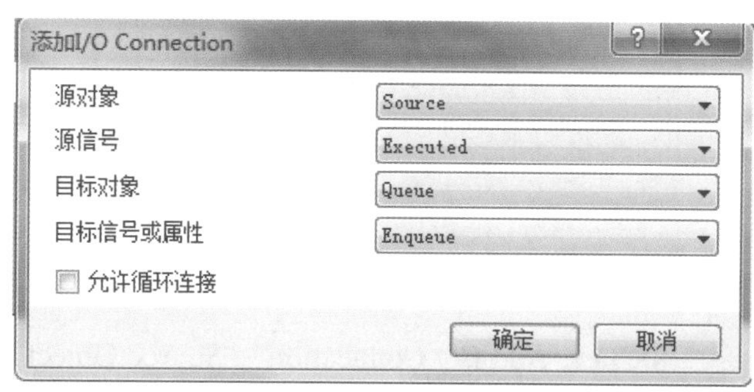

Fig.15-44 Add the Entry Signal Link of the Copied Object

图15-44 添加拷贝对象入列信号连结

（6）点击“添加 I/O Connection”，“源对象”选择“PlaneSensor”，“源信号”选择“SensorOut”，“目标对象”选择“Queue”，“目标信号或属性”选择“Dequeue”，如图15-45所示。

(6) Click “AddI/OConnection”, select “Plane Sensor” for “Source Object”, select “SensorOut” for “Source Signal”, select “Queue” for “Target Object”, and select “Dequeue” for “Target Signal or Property” 1, as shown in Fig. 15-45.

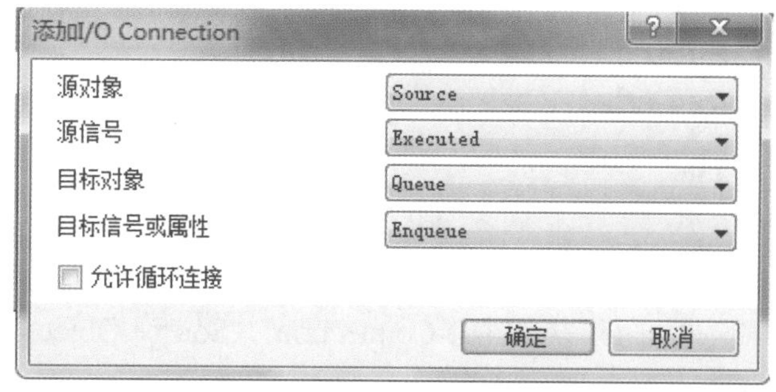

Fig.15-45 Add the Entry Signal Link of the Copied Object

图15-45 添加拷贝对象出列信号连结

（7）点击“组成”，点击“添加组件”，点击“信号和属性”，选择“LogicGate”，鼠标右键点击“Logic Gate”，点击“属性”，“Operator”设置为“NOT”，点击“关闭”，如图15-46所示。

(7)Click “Composition”, click “Add component”, click “Signals and Attributes”, select “Logic Gate”, right-click “Logic Gate”, click “Properties”, “Operator” is set to “NOT”, and click “Close”, as shown in Fig. 15-46.

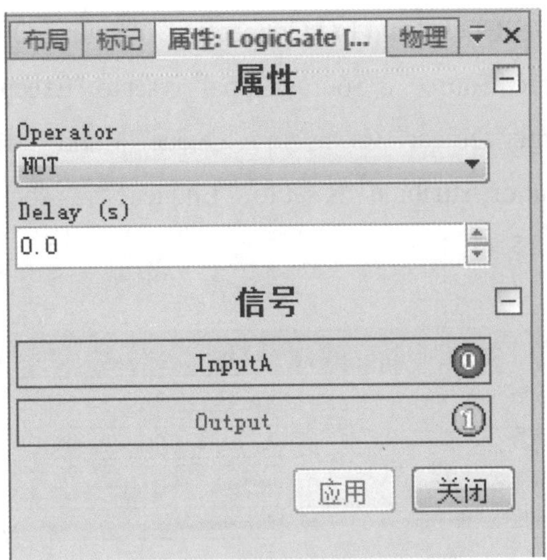

Fig.15-46 Adding NOT Logic Components

图 15-46 添加 NOT 逻辑组件

(8)Click "Add I/O Connection", "Source Object" is set to "PlaneSensor", "Source Signal" is set to "SensorOut", "Target Object" is set to "LogicGate [NOT]", "Target Signal or Attribute" is set to "Input A", and click OK, as shown in Fig. 15-47.

（8）点击"添加 I/O Connection"，"源对象"设置为"PlaneSensor"，"源信号"设置为"SensorOut"，"目标对象"设置为"LogicGate【NOT】"，"目标信号或属性"设置为"Input A"，点击确定，如图 15-47 所示。

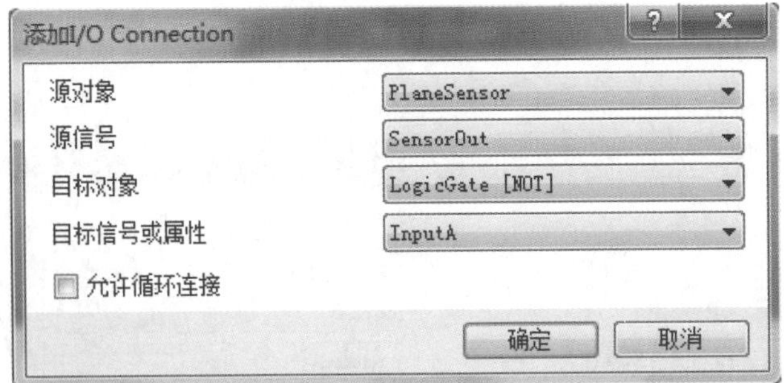

Fig.15-47 Add the Inverted Signal Link of the Sensor Output

图 15-47 添加传感器输出取反的信号连结

(9)Click "Add I/O Connection", "Source Object" is set to "LogicGate[NOT]", "Source Signal" is set to "Output", "Target Object" is set to "Source", "Target Signal or Attribute" is set to "Execute", and click "OK", as shown in Fig. 15-48.

（9）点击"添加 I/O Connection"，"源对象"设置为"LogicGate【NOT】"，"源信号"设置为"Output"，"目标对象"设置为"Source"，"目标信号或属性"设置为"Execute"，点击"确定"，如图 15-48 所示。

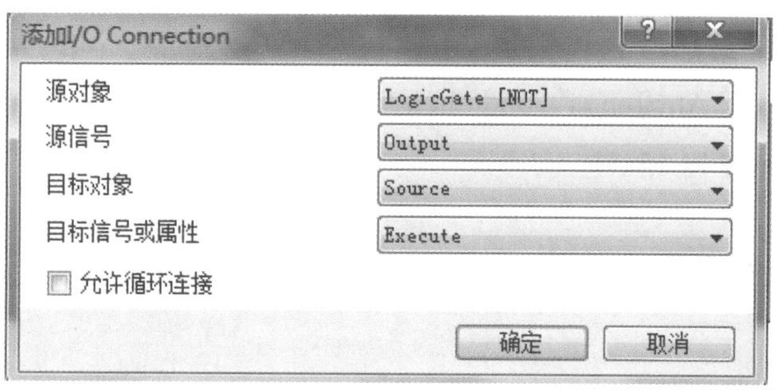

Fig.15-48 Trigger Source to Run

图 15-48 触发 Source 运行

（10）点击“信号和连结”，点击“添加 I/O Connection”，“源对象”设置为“PlaneSensor”，“源信号”设置为“SensorOut”，“目标对象”设置为“SC_Infeeder”，“目标信号或属性”设置为“BoxInPos”，点击“确定”，如图 15-49 所示。

(10) Click “Signal and Connection”, click “Add I/O Connection”, “Source Object” is set to “PlaneSensor”, “Source Signal” is set to “SensorOut”, “Target Object” is set to “SC_Infeeder”, “Target Signal or Attribute” is set to “BoxInPos”, and click “OK”, as shown in Fig. 15-49.

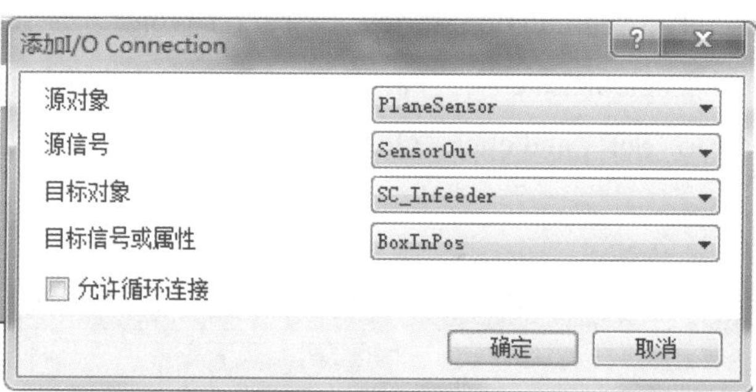

Fig.15-49 Add the Signal Link of the Sensor Output In-position Signal

图 15-49 添加传感器输出到位信号的信号连结

（11）在“组成”里点击“添加组件”，点击“其他”，点击“SimulationEvents”，如图 15-50 所示。

(11) Click “Add Component” in “Composition”, click “Other” ,and click “SimulationEvents”, as shown in Fig. 15-50.

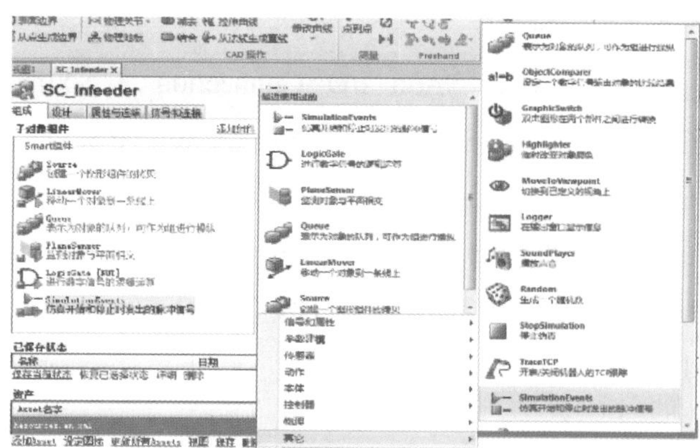

Fig. 15-50 Calling the SimulationEvents Component

图 15-50 调用 SimulationEvents 组件

(12) Click "Add Component", click "Signals and Attributes", and click "LogicSRLatch", as shown in Fig.15-51.

（12）点击"添加组件"，点击"信号和属性"，点击"LogicSRLatch"，如图 15-51 所示。

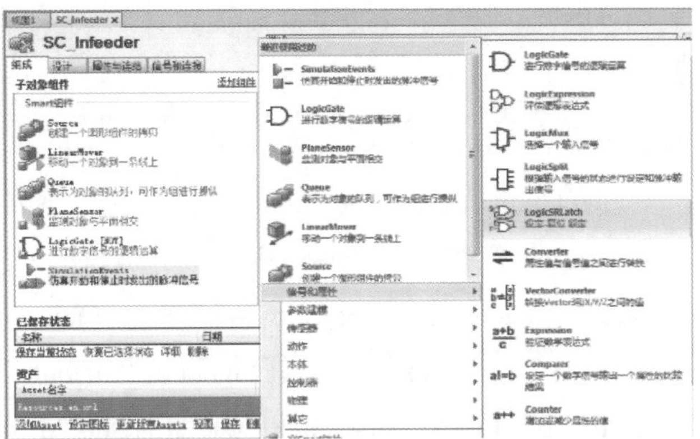

Fig.15-51 Calling the LogicSRLatch Component

图 15-51 调用 LogicSRLatch 组件

(13) Click "Add I/O Connection", "Source Object" is set to "SimulationEvents", "Source Signal" is set to "SimulationStarted", "Target Object" is set to "LogicSRLatch", "Target Signal or Attribute" is set to "Set", and click "OK", as shown in Fig. 15-52.

（13）点击"添加 I/O Connection"，"源对象"设置为"SimulationEvents"，"源信号"设置为"SimulationStarted"，"目标对象"设置为"LogicSRLatch"，"目标信号或属性"设为"Set"，点击"确定"，如图 15-52 所示。

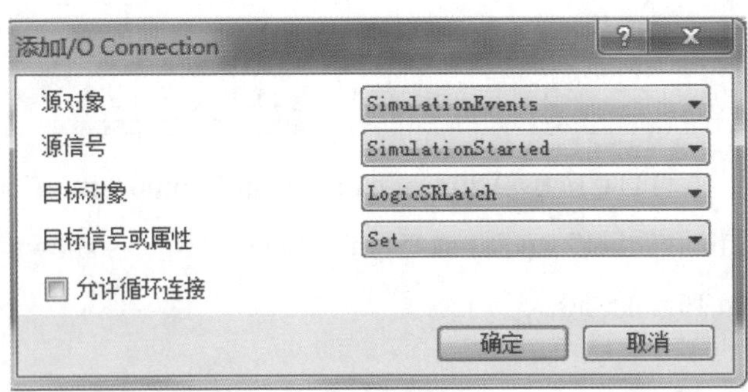

Fig.15-52 SimulationStarted Signal Sets LogicSRLatch

图 15-52 SimulationStarted 信号置位 LogicSRLatch

(14) Click "Add I/O Connection", "Source Object" is set to "SimulationEvents", "Source Signal" is set to "SimulationStopped", "Target Object" is set to "LogicSRLatch", "Target Signal or Attribute" is set to " Reset", and click "OK", as shown in Fig. 15-53.

（14）点击"添加 I/O Connection"，"源对象"设置为"SimulationEvents"，"源信号"设置为"SimulationStopped"，"目标对象"设置为"LogicSRLatch"，"目标信号或属性"设置为"Reset"，点击"确定"，如图 15-53 所示。

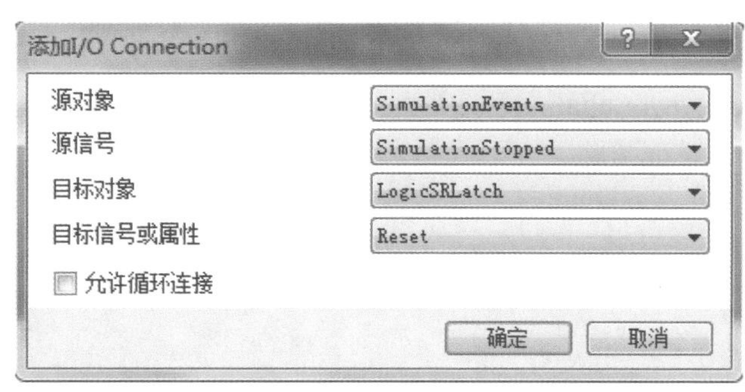

Fig. 15-53　SimulationStopped Signal Resets LogicSRLatch
图 15-53　SimulationStopped 信号复位 LogicSRLatch

（15）点击“添加 I/O Connection”，“源对象”设置为“LogicSRLatch”，“源信号”设置为“Output”，“目标对象”设置为“PlaneSensor”，“目标信号或属性”设置为“Active”，点击“确定”，图如 15-54。

(15) Click “Add I/O Connection”, “Source Object” is set to “LogicSRLatch”, “Source Signal” is set to “Output”, “Target Object” is set to “Plane Sensor”, “Target Signal or Attribute” is set to “ Active”, and click OK, as shown in Fig. 15-54.

Fig. 15-54　LogicSRLatch Output Signal Activates PlaneSensor
图 15-54　LogicSRLatch 输出信号激活 PlaneSensor

（16）点击“仿真”，点击“仿真设定”，取消勾选机器人系统，点击“关闭”，如图 15-55 所示。

（17）回到“view1”，点击“播放”，鼠标右键点击创建的 SMART 组件，手动触发一下启动信号 StartCNV 信号。这时我们可以看到物料被拷贝，并沿着输送链移动，一直运动到

(16)Click“Simulation”,click “Simulation Settings”, uncheck the robot system, and click “Close”, as shown in Fig. 15-55.

(17) Back to “view1”, click “Play”, right-click the created SMART component, and trigger the StartCNV signal manually. At this time, we can see that the material is copied and moved along the conveyor chain until it stops at the end of the conveyor chain, and the in-position signal is set to 1.

输送链末端停下，并且，到位信号设置为1。

(18) Remove the material at the end of the conveyor chain to see if the material can be copied for the next time.

(18)移开输送链末端的物料，看能否使物料进行下一次拷贝。

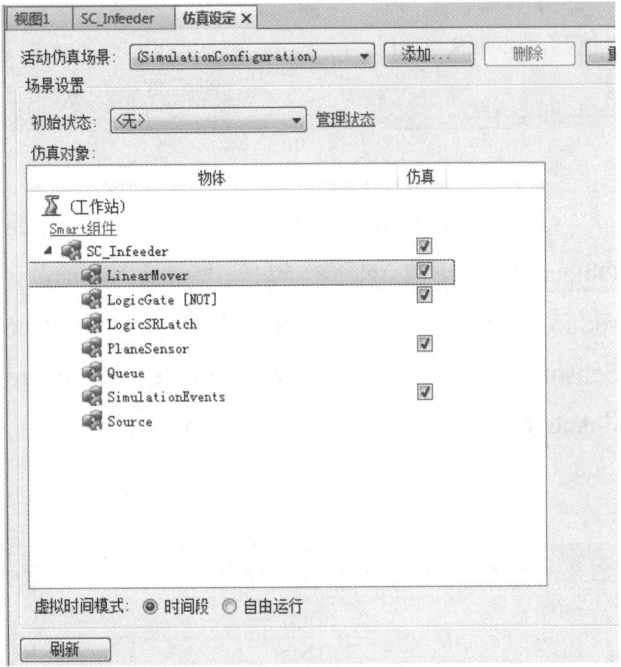

Fig. 15-55 Simulation Settings

图 15-55 仿真设定

REFERENCES
参考文献

［1］姚屏. 工业机器人技术基础［M］. 北京：机械工业出版社，2020.

［2］田小静. 工业机器人技术基础及应用［M］. 北京：机械工业出版社，2020.

［3］蒋正炎，陈永平，汤晓华. 工业机器人应用技术［M］. 北京：高等教育出版社，2021.

［4］温宏愿，孙松丽，林燕文. 工业机器人技术及应用［M］. 北京：高等教育出版社，2019.

［5］韩鸿鸾. 工业机器人现场编程与调试一体化教程［M］. 西安：西安电子科技大学出版社，2021.

［6］张郭. 工业机器人编程技术及实践［M］. 西安：西安电子科技大学出版社，2020.

［7］吕世霞，周宇，沈玲. 工业机器人现场操作与编程［M］. 武汉：华中科技大学出版社，2021.

［8］宋永端. 工业机器人系统及其先进控制方法［M］. 北京：科学出版社，2019.

［9］周宇，范俐. ABB 工业机器人典型工程应用［M］. 西安：西安电子科技大学出版社，2020.

［10］叶泉，朱红娟，孟庆波. 工业机器人操作与编程［M］. 北京：机械工业出版社，2021.